Student Study Guide

to accompany

Hole's Essentials of Human Anatomy & Physiology

Eleventh Edition

David Shier
Washtenaw Community College

Jackie Butler
Grayson County College
Department of Science & Mathematics

Ricki Lewis
Alden March Bioethics Institute

Prepared by

Nancy A. Sickles Corbett, Ed. D., NP-C, RN.
Family Nurse Practitioner, Primary Care of Moorestown Moorestown, NJ

Contribution by

Patrice Parsons
Grayson County College

The McGraw-Hill Companies

Student Study Guide to accompany
HOLE'S ESSENTIALS OF HUMAN ANATOMY & PHYSIOLOGY, ELEVENTH EDITION
DAVID SHIER, JACKIE BUTLER, AND RICKI LEWIS

Published by McGraw-Hill Higher Education, an imprint of The McGraw-Hill Companies, Inc., 1221 Avenue of the Americas, New York, NY 10020. Copyright © 2012, 2009, 2006 and 2003 by The McGraw-Hill Companies, Inc. All rights reserved.

No part of this publication may be reproduced or distributed in any form or by any means, or stored in a database or retrieval system, without the prior written consent of The McGraw-Hill Companies, Inc., including, but not limited to, network or other electronic storage or transmission, or broadcast for distance learning.

This book is printed on acid-free paper.

1 2 3 4 5 6 7 8 9 0 QDB/QDB 1 0 9 8 7 6 5 4 3 2 1

ISBN: 978-0-07-733884-8
MHID: 0-07-733884-7

om

CONTENTS

TO THE STUDENT

A study guide attempts to do what the name implies: guide your study so that your learning efforts are more efficient. This study guide is based on several beliefs: (1) learning occurs best when the learner is active rather than passive; (2) learning is easiest when the material is organized in simple units; and (3) the learner can best evaluate what he or she knows well, is unsure of, and does not know.

The study guide chapters correspond to the chapters in Hole's Essentials of Human Anatomy and Physiology, 11th edition, by David Shier, Jackie Butler, and Ricki Lewis. The elements of the study guide chapters and their purposes are described below.

1. Overview. The learning outcomes at the beginning of each chapter in the text are arranged in groups according to broad, general concepts presented in the chapter. The overview also contains a purpose statement that offers a rationale for studying the chapter.
2. Chapter Outcomes. The chapter learning outcomes from the text are listed to help guide your study of the chapter.
3. Focus Question. The focus question helps you focus your study of each chapter.
4. Mastery Test. The mastery test, taken before reading the chapter, is designed to help you identify
   a. concepts you already know.
   b. concepts you need to clarify.
   c. concepts you do not know.

   If you are using a study guide for the first time, you may be unfamiliar with this type of testing. It is important for you to realize that this test is for your information. Its purpose is to help you learn where to concentrate your learning efforts; therefore, it is best not to guess at any answers. There are a variety of types of questions in the Mastery Tests including true/false, fill in the blanks, matching, and multiple choice questions. It is important for you to know that some of the multiple choice questions may have more than one correct response choice. If there is more than one correct choice, you must choose each correct response in order to say you have mastered the material. If you fail to identify all the correct choices, you have identified an area on which you may want to concentrate further study.

5. Study Activities. A variety of study activities help facilitate the study of the principal ideas of each chapter. The study activities should be done after you have read the chapter carefully, concentrating on those areas that the mastery test indicates you do not know.

   The first activity in each unit is a vocabulary exercise, concentrating on word parts appropriate to each chapter. You are asked to define these as you understand them and then compare your definitions with those in the chapter. You may find it helpful to define terms orally and in writing. If you have a tape recorder, you may wish to use it as a study device.

   After the vocabulary exercise, you may be asked to describe a process, label a diagram, fill in a table, or observe the function of a body part in yourself or in a partner. (This partner may be a classmate or a family member.) These are written activities, but you may also find it helpful to repeat them orally.

   After you complete the study activities, retake the mastery test. A comparison of the two scores will indicate the progress you have made. You may also wish to set a learning goal for yourself, such as a score of 70 percent, 80 percent, or 90 percent on the mastery test after completing your study of a chapter. If you have not attained your goal, the mastery test results can show where you need additional study.

   The answers to the mastery test are at the end of the study guide. You can compare your responses to the review activities by referring to the appropriate page numbers in the text. Each major section in the study guide is identified by a Roman numeral and the title of the corresponding section in the text. The activities in the study guide are lettered, and the corresponding pages in the text are noted after the activity. In addition, the Study Activity answers are available in an answer for each chapter.

You are responsible for your own learning; no teacher can assume that responsibility. A study guide can help you direct your study more efficiently, but only you can control how well and how completely you use the guide.

UNIT I

LEVELS OF ORGANIZATION

CHAPTER 1

INTRODUCTION TO HUMAN ANATOMY AND PHYSIOLOGY

OVERVIEW

This chapter begins the study of anatomy and physiology by defining the disciplines (learning outcome 2) and by explaining the characteristics and needs that are common to all living things (learning outcomes 3, 4, and 6). It introduces a basic mechanism necessary to maintain life (learning outcome 7), as well as the relationship of increasingly complex levels of organization in humans (learning outcome 3). The study of levels of organization continues with the identification of body cavities and the organs to be found within each cavity (learning outcomes 9–12). Finally, the language used to describe relative positions of body parts, body sections, and body regions is presented (learning outcome 14) and the mechanisms of homeostasis are addressed (learning outcome 8).

This chapter defines the characteristics and needs common to all living things and the manner in which the human body is organized to accomplish life processes. The language particular to anatomy and physiology is also introduced.

LEARNING OUTCOMES

After you have studied this chapter, you should be able to

1.1 Introduction

1. Identify some of the early discoveries that led to our understanding of the body.

1.2 Anatomy and Physiology

2. Explain how anatomy and physiology are related.

1.3 Levels of Organization

3. List the levels of organization in the human body and the characteristics of each.

1.4 Characteristics of Life

4. List and describe the major characteristics of life.
5. Give examples of metabolism.

1.5 Maintenance of Life

6. List and describe the major requirements of organisms.
7. Explain the importance of homeostasis to survival.
8. Describe the parts of a homeostatic mechanism and explain how they function together.

1.6 Organization of the Human Body

9. Identify the locations of the major body cavities.
10. List the organs located in each major body cavity.
11. Name and identify the locations of the membranes associated with the thoracic and abdominopelvic cavities.
12. Name the major organ systems, and list the organs associated with each.
13. Describe the general functions of each organ system.

1.7 Anatomical Terminology

14. Properly use the terms that describe relative positions, body sections, and body regions.

FOCUS QUESTION

How does knowledge of anatomy and physiology facilitate communication about body structure and function between scientists and health-care professionals?

MASTERY TEST

Now take the mastery test. Do not guess. As soon as you complete the test, correct it. Note your successes and failures so that you can read the chapter to achieve your learning outcomes.

1. Study of the human body first began with earliest humans because
   a. our early ancestors were curious about the world around them.
   b. they were as interested in their body parts and their functions as we are today.
   c. of their concern with illness and injury.
2. Which of the following factors sets the stage for early knowledge of the human body?
   a. a belief that spirits or gods controlled sickness and health
   b. the growing experience of medicine men as they treated the sick with herbs and potions
   c. the development of paper
   d. the ability to ask health questions and record the answers
3. The development of modern science began with
   a. rejection of the belief in supernatural forces.
   b. the growing experience of medicine men in the treatment of illness and injury.
   c. the belief that natural processes were caused by forces that could be understood.
   d. the ability to ask questions and record the answers.
4. What languages form the basis of the language of anatomy and physiology?
5. The branch of science that deals with the structure of body parts is ________________________.
6. The branch of science that studies how body parts function is________________________.
7. The function of a part is (always, sometimes, never) related to its structure.
8. Whereas knowledge of physiology continues to develop, knowledge of anatomy does not change.
   a. True
   b. False
9. List the levels of organization of the body in order of increasing complexity, beginning with the atom.
10. The sum total of chemical reactions in the body that break substances down and build them up is __________________.
11. List those characteristics that humans share with other organisms.
    a.
    b.
    c.
    d.
    e.
    f.
    g.
    h.
    i.
    j.
12. The most abundant chemical in the human body is______________________________.
13. Food is used as a(n) ____________source to build new _____________and to participate in the regulation of chemical reactions.
14. Oxygen is used to release____________________________.
15. An increase in temperature ____________________________ the rate of chemical reactions.
16. The action of the heart creates _________________ pressure in the blood vessels.
17. Homeostasis means
    a. maintenance of a stable internal environment.
    b. integrating the functions of the various organ systems.
    c. preventing any change in the organism.

18. Match the terms related to homeostasis in column A to the definitions in column B.

a. homeostasis

b. receptors

c. effectors

d. set point

1. ___ point(s) that tell(s) what a particular value should be

2. ___ provide(s) information about specific conditions in the internal environment

3. ___ cause(s) responses that alter conditions in the internal environment

19. The portion of the body that contains the head, neck, and trunk is called the ______________________.

20. The arms and legs are called the______________ portion.

21. The major cavities of the axial portion of the body are the ______________ cavity, the ______________ canal, the ______________ cavity, and the ______________cavity.

22. The inferior boundary of the thoracic cavity is the ______________________ .

23. The heart, esophagus, trachea, and thymus gland are located in the______________________ , which separates the thoracic cavity into two compartments.

24. The pelvic cavity is

a. the portion of the abdominopelvic cavity below the pelvic brim.

b. the portion of the abdomen that contains the internal reproductive organs and the urinary bladder.

c. the portion of the abdomen surrounded by the bones of the pelvis.

25. List the four body cavities located in the head.

26. The visceral and parietal pleural membranes secrete a serous fluid into a potential space called the______________________.

27. The heart is covered by the ______________________membranes.

28. The peritoneal membranes are located in the ______________________ cavity.

29. The covering of the body is made of an organ and various accessory organs known as the______________________ system.

30. Match the systems listed in the first column with the functions listed in the second column.

____a. nervous system
____b. muscular system
____c. circulatory system
____d. respiratory system
____e. skeletal system
____f. digestive system
____g. lymphatic system
____h. endocrine system
____i. urinary system
____j. reproductive system

1. reproduction
2. processing and transporting
3. integration and coordination
4. support and movement

31. Which of the following positions of body parts is (are) in anatomical position?

a. palms of hands turned toward sides of body

b. standing erect

c. arms at sides

d. face toward left shoulder

32. Terms of relative position are used to describe the

a. relationship of siblings within a family.

b. importance of the various functions of organ systems in maintaining life.

c. location of one body part with respect to another.

33. A sagittal section divides the body into

a. superior and inferior portions.

b. right and left portions.

c. anterior and posterior portions.

34. The terms *epigastric, hypochondriac,* and *iliac* are used to describe what?

STUDY ACTIVITIES

Aids to Understanding Words

Define the following word parts. (p. 2)

append-

cardi-

cran-

dors-

homeo-

-logy

meta-

pariet-

pelv-

peri-

pleur-

-stasis

-tomy

1.1 Introduction (p. 2)

A. How did ancient healers begin their study of the human body?

B. How did the scientific study of the human body begin?

1.2 Anatomy and Physiology (p. 3)

A. Explain how the structure of the fingers is related to their grasping function.

B. Are new discoveries more likely in anatomy or in physiology? Explain your answer.

1.3 Levels of Organization (pp. 3–4)

Arrange the following structures in increasing levels of complexity: atoms, organ systems, organelles, organism, organs, macromolecules, cells, tissues, molecules.

1.4 Characteristics of Life (pp. 4–5)

A. Describe the following characteristics of life: (pp. 4–5)

movement

responsiveness

growth

reproduction

respiration

digestion

absorption

assimilation

circulation

excretion

B. What is metabolism? (p. 4)

1.5 Maintenance of Life (pp. 5–8)

A. Match the terms in the first column with the statements in the second column that define their role in the maintenance of life.

____a. water	1. essential for metabolic processes
____b. food	2. governs the rate of chemical reactions
____c. oxygen	3. creates a pressing or compressing action
____d. heat	4. necessary for release of energy
____e. pressure	5. provides chemicals for building new living matter

B. Why are observations of the vital signs important to nurses and physicians? (p. 5)

C. Answer these questions concerning homeostasis. (pp. 5–8)

1. Define *homeostasis*. Include the functions of receptors, effectors, and a set point.

2. How is body temperature maintained at 37°C (98.6°F)?

3. Describe negative and positive feedback mechanisms. Give examples of each.

1.6 Organization of the Human Body (pp. 8–14)

A. List the components of the axial and appendicular portions. (p. 8)

B. 1. List the contents of the thoracic cavity. (pp. 8–9)

2. List the contents of the abdominopelvic cavity. (pp. 8–9)

C. List the four smaller cavities within the head. (pp. 8–10)

D. Answer these questions concerning the thoracic and abdominopelvic membranes. (pp. 10–11)

1. Fill in the blanks.
   a. The walls of the thoracic cavity are lined with a membrane called the____________________.
   b. The lungs are covered by the ____________________.
   c. Why is the pleural cavity called a potential space?

2. Name and describe the membranes covering the heart.

3. The linings of the abdominopelvic cavity are the____________________and the____________________.

E. Fill in the following table concerning the structure and function of organ systems. (pp. 12–14)

Function	Organ system	Function
Support and movement	1.	
	2.	
Integration and coordination	1.	
	2.	
Transport	1.	
	2.	
	3.	
Absorption and excretion	1.	
	2.	
	3.	
Reproduction: Female	1.	
Male	2.	

1.7 Anatomical Terminology (pp. 14–17)

A. Using this illustration, specify the terms that describe the relationship of one point on the body to another. (p. 14)

1. Point *a* in relation to point *d*

2. Point *f* in relation to point *h*

3. Point *g* in relation to point *i*

4. Point *l* in relation to point *j*

5. Point *i* in relation to point *g*

6. Point *c* in relation to point *a*

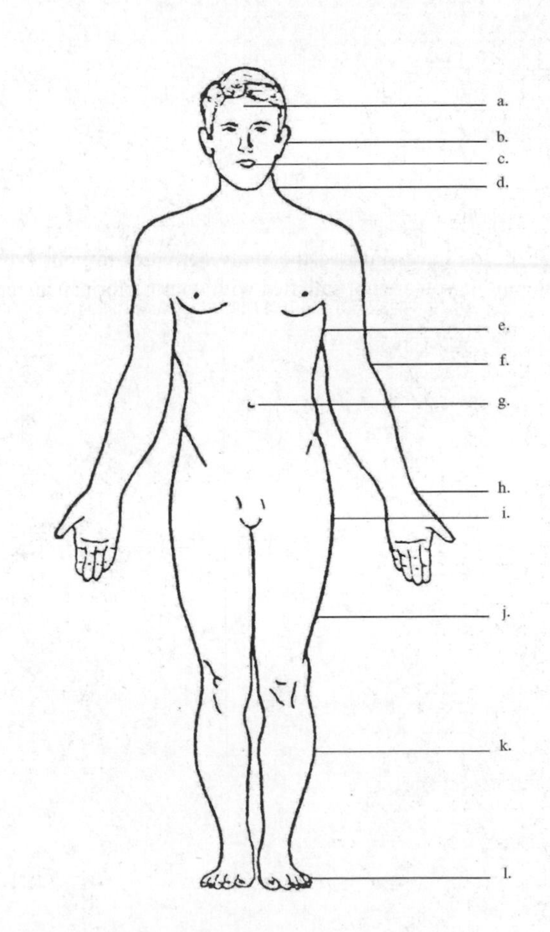

B. Using the illustration on the previous page, complete the following exercises. (pp. 14–17)

1. Draw a line through the drawing to indicate a midsagittal section. How is this different from a frontal section?
2. Draw a line through the drawing to indicate a transverse section.
3. Define *cross section, longitudinal section*, and *oblique section.*
4. Locate and label the following body regions on the diagram: epigastric, umbilical, hypogastric, hypochondriac, lumbar, iliac. Locate these regions on yourself or on a partner.
5. Locate and label the following body parts on the diagram: antebrachium, antecubital, axillary, brachial, buccal, cervical, groin, inguinal, mammary, ophthalmic, palmar, pectoral.

Clinical Focus Question

List the organs and/or systems whose structure and function should be assessed to diagnose the cause of the following symptoms.

A. An earache and pressure behind the eyes

B. Vomiting and diarrhea

C. Chest pain

D. Low back pain

When you have finished the study activities to your satisfaction, retake the mastery test and compare your results with your initial attempt. If you are not satisfied with your performance, repeat the appropriate study activities.

CHAPTER 2

CHEMICAL BASIS OF LIFE

OVERVIEW

This chapter introduces some basic concepts of chemistry, a science that studies the composition of substances and the changes that occur as basic elements combine. It explains how substances combine to make up matter (learning outcomes 1–5), how substances are classified as an acid or a base (learning outcomes 6 and 7), and the organic and inorganic substances that make up the living cell (learning outcomes 8 and 9).

Knowledge of basic chemical concepts enhances understanding of the functions of cells and of the human body.

LEARNING OUTCOMES

After you have studied this chapter, you should be able to

2.1 Introduction

1. Give examples of how the study of living material requires an understanding of chemistry.

2.2 Structure of Matter

2. Describe how atomic structure determines how atoms interact.
3. Describe the relationships among matter, atoms, and molecules.
4. Explain how molecular and structural formulas symbolize the composition of compounds.
5. Describe three types of chemical reactions.
6. Explain what acids, bases, and buffers are.
7. Define *pH* and be able to use the pH scale.

2.3 Chemical Constituents of Cells

8. List the major inorganic chemicals common in cells and identify the functions of each.
9. Describe the general functions of the four main organic chemicals in cells.

FOCUS QUESTION

How is chemistry related to the structure and function of living things and their parts?

MASTERY TEST

Now take the mastery test. Do not guess. As soon as you complete the test, correct it. Note your successes and failures so that you can read the chapter to meet your learning needs.

1. The human body is composed of chemicals.

   a. True
   b. False

2. What is matter? In what forms can it be found?

3. The substances that constitute all matter are called ______________________.

4. What four elements are most plentiful in the human body?

5. An atom is made of

   a. a nucleus.
   b. protons.
   c. neutrons.
   d. electrons.
   e. all of the above.

6. Match the following.

____a. neutron
____b. proton
____c. electron

1. positive electrical charge
2. negative electrical charge
3. no electrical charge

7. The atomic number of an element is determined by the number of ________________________.

8. When atoms combine, they gain or lose
a. electrons.
b. neutrons.
c. protons.
d. nuclei.

9. Atoms that have the same atomic numbers but different atomic weights are
a. catalysts.
b. neutrinos.
c. isotopes.
d. heliotropes.

10. The atomic weight of an element is determined by the number of ________________________.

11. Why would an element be considered inactive or inert?

12. An ion is
a. an atom that is electrically charged.
b. an atom that has gained an electron.
c. an atom that has lost an electron.
d. all of the above.

13. An ionic bond is created by
a. a positive and a negative ion attracting each other.
b. two or more positive ions combining.
c. two or more negative ions combining.

14. In forming a covalent bond, electrons are
a. shared by two atoms.
b. given up by an atom.
c. taken up by an atom.
d. none of the above.

15. Water is formed by molecules of hydrogen and oxygen united by a __________________ bond.

16. A compound is formed when atoms of __________________ elements combine.

17. $C_6H_{12}O_6$ is an example of a(n) __________________ formula.

18. This figure is an example of a(n) __________________formula.

19. Two major types of chemical reactions are called ________________and ________________.

20. The symbol ⇄ indicates a(n) ______________________ reaction.

21. An atom or a molecule that affects the rate of a reaction without being consumed by the reaction is called a(n) ________________.

22. An electrolyte that releases hydrogen ions in water is a(n) ______________________.

23. Electrolytes that release ions that combine with hydrogen ions are called ______________________.

24. The pH measures concentration of ______________________.

25. What is the pH of a neutral solution?

26. An inorganic substance that releases ions when it reacts with water is known as a(n)______________________.

27. Identify the following cell constituents with an *O* if they are organic and an *I* if they are inorganic.

a. water ( )
b. carbohydrate ( )
c. glucose ( )
d. oxygen ( )
e. protein ( )
f. fats ( )

28. Carbohydrate molecules contain atoms of ____________________, ____________________, and____________________.

29. Fat molecules contain *mostly* ____________________ and ____________________.

30. Fats are used in the body primarily to store____________________.

31. Fats, phospholipids, and steroids are important____________________ found in the human cell.

32. An enzyme is a(n) ____________________that acts as a catalyst.

33. In addition to carbon, hydrogen, and oxygen, proteins also contain atoms of____________________.

34. What characteristic of protein determines its function?

35. The two types of nucleic acids are____________________ and ____________________.

36. The function of nucleic acids is to
a. store information and control life processes.
b. act as receptors for hydrogen ions.
c. neutralize bases within the cell.

STUDY ACTIVITIES

Aids to Understanding Words

Define the following word parts. (p. 31)

di-

glyc-

lip-

-lyt

mono-

poly-

sacchar-

syn-

2.1 Introduction (p. 31)

A. What does the study of chemistry involve?

B. Why is chemistry essential to understanding body structure and function?

C. List some of the chemicals found in the human body.

2.2 Structure of Matter (pp. 31–39)

A. Answer these questions concerning elements and atoms. (pp. 31–32)

1. Anything that has weight and takes up space is____________________.
2. Fundamental substances are called ____________________.
3. Tiny, invisible particles that comprise fundamental substances are called ____________________.
4. Two or more particles of fundamental substance can combine forming a(n) ____________________.
5. The particles found in the nucleus of an atom are ____________________ and ____________________.
6. The particle that moves around the nucleus of an atom and carries a negative electrical charge is the ____________________.
7. How many elements are required by living organisms?

B. Supply the missing elements or symbols in the table. (p. 32)

Element	Symbol	Element	Symbol
Oxygen		Sodium	
Carbon		Magnesium	
	H	Cobalt	
Nitrogen			Cu
	Ca		F
	P	Iodine	
	K		Fe
Sulfur			Mn
	Cl	Zinc	

C. Answer these questions that pertain to the accompanying illustration. (pp. 32–35)

1. Atoms that become electrically charged by gaining or losing an electron are called ____________________.

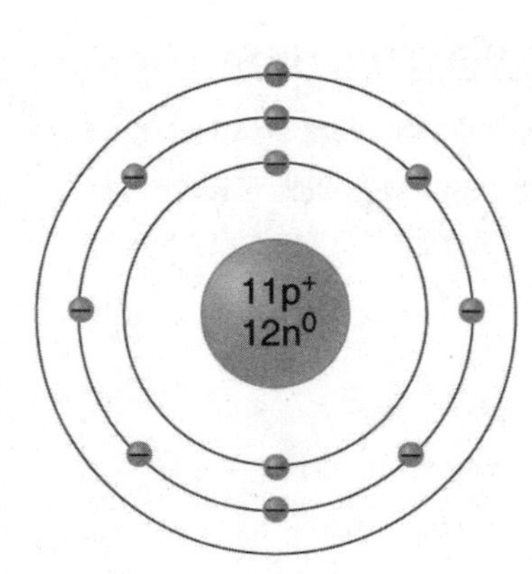

2. What is the atomic number of this atom?

3. How many electrons are needed to fill its outer shell?

4. Is this element active or inert?

5. Identify this element.

D. 1. Diagram an ionic bond (electrovalent bond) and a covalent bond. (pp. 33–36)

2. How are these bonds different?

3. Describe a polar molecule and its hydrogen bond.

E. Answer the following questions about radioactive isotopes. (pp. 33–34)

1. What is an isotope?

2. Describe the difference between a stable isotope and a radioactive isotope.

3. Why are isotopes useful in the treatment of cancer?

F. Answer these questions that pertain to the accompanying diagram. (pp. 37–39)

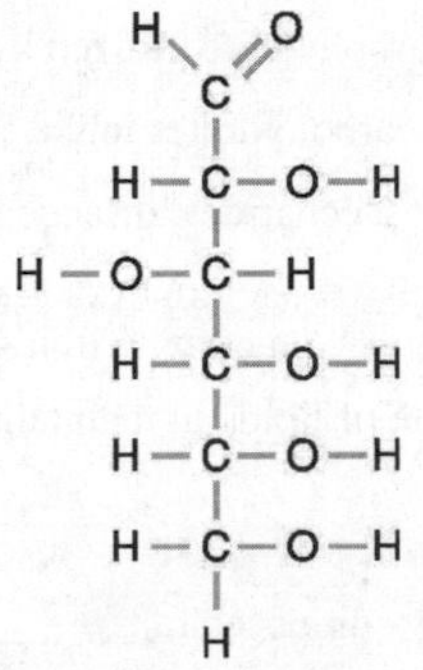

1. Give the molecular formula of this substance.

2. Identify this compound.

3. Is it an organic or an inorganic substance? Why?

```
        H   O
         \ //
          C
          |
      H - C - O - H
          |
  H - O - C - H
          |
      H - C - O - H
          |
      H - C - O - H
          |
      H - C - O - H
          |
          H
```

G. Label the following chemical reactions. (p. 38)

1. $A + B \rightarrow AB$

2. $AB \rightarrow A + B$

3. $A + B \rightleftarrows AB$

H. Answer these questions concerning acid and base concentrations. (pp. 38–39)

1. What is meant by pH?

2. Substances that release hydrogen ions (H^+) in water are called____________________.
3. Substances that release ions that then combine with hydrogen ions are called____________________.
4. Identify these substances as either acid or base.

carrot, pH 5.0: tomato, pH 4.2:
milk of magnesia, pH 10.5: lemon, pH 2.3:
human blood, pH 7.4 : distilled water, pH 7.0:

5. What is the range of pH within which the human organism can survive?

6. A chemical that resists a change in pH is a ____________________.

2.3 Chemical Constituents of Cells (pp. 39–46)

A. What roles do the following inorganic substances play in the cell? Be specific. (p. 40)

water

carbon dioxide

oxygen

inorganic salts (Na^+, K^+, Ca^{+2}, HCO_3^-, PO_4^{-3})

B. Answer these questions concerning carbohydrates. (pp. 41–42)

1. What is the role of carbohydrates in maintaining the cell?
2. The form of carbohydrate utilized by the cell is____________________.
3. Humans store carbohydrates in the form of ___________________________.
4. Describe monosaccharides, disaccharides, and polysaccharides.

C. Answer these questions concerning lipids. (pp. 42–43)

1. What is the role of lipids in maintaining the cell?
2. Fats are *mostly* composed of ____________________ and______________________.
3. Fats containing single carbon—carbon bonds are_____________________________.
4. Fats containing one or more double-bonded carbon atoms are___________________________.
5. Describe the molecular structure of these lipids.

   triglycerides

   phospholipids

   steroids

6. Identify the characteristics and functions of these lipids.

   triglycerides

   phospholipids

   steroids

D. Answer these questions concerning proteins. (pp. 43–45)

1. What is the role of protein in maintaining the cell?

2. The building blocks of a protein are____________________.

3. The shape or conformation of protein molecules is maintained by its____________________________.

4. A protein molecule that has become disorganized and lost its shape is said to be______________________.

5. How does protein conformation affect its function?

E. Answer these questions concerning nucleic acids. (pp. 44–47)

1. The two types of nucleic acids are ______________________ and _________________________.

2. What is their role in cell function?

Clinical Focus Questions

A. Why is it important for human beings to have an adequate amount of water?

B. What is likely to happen to chemical activity if the body is deprived of adequate amounts of water?

C. List mechanisms by which an individual may become deficient in water.

D. How do prions cause disease?

When you have finished the study activities to your satisfaction, retake the mastery test and compare your results with your initial attempt. If you are not satisfied with your performance, repeat the appropriate study activities.

CHAPTER 3

CELLS

OVERVIEW

This chapter presents the structural unit of the body—the cell. It describes what a cell is and how cells within the human body vary (learning outcome 1). Because cells vary so widely, a composite cell is used to study cells. Learning outcomes 2 through 4 ask you to identify the major parts of a component cell, the functions of organelles within the cytoplasm, and the structure and function of the cell membrane; and they prepare you to begin the study of genetics by being able to describe the parts of the cell nucleus and their function. Achievement of learning outcome 5 allows you to explain how different substances move through the cell membrane, as well as how the mechanisms used by various substances to cross the cell membrane differ. Learning outcomes 6 and 7 focus on understanding the reproduction of cells and the cell cycle. Lastly, learning outcomes 8 and 9 provide an understanding of determination and the fates of cells.

LEARNING OUTCOMES

After you have studied this chapter, you should be able to

3.1 Introduction

1. Explain how cells differ from one another.

3.2 Composite Cell

2. Explain how the structure of a cell membrane makes possible its functions.
3. Describe each type of organelle, and explain its function.
4. Describe the parts of the cell nucleus and their functions.

3.3 Movements Through Cell Membranes

5. Explain how substances move into and out of cells.

3.4 The Cell Cycle

6. Explain why regulation of the cell cycle is important to health.
7. Describe the cell cycle.
8. Explain how stem cells and progenitor cells make possible growth and repair of tissues.
9. Explain how two differentiated cell types can have the same genetic information but different appearances and functions.

FOCUS QUESTION

How does the structure of cellular organelles contribute to and support the functions of the organelle and, in turn, the cell?

MASTERY TEST

Now take the mastery test. Do not guess. As soon as you complete the test, correct it. Note your successes and failures so that you can read the chapter to meet your learning needs.

1. The cells of the human body vary in _______________ and _______________ of component structures. The function of various cells is made possible by_______________________________.
2. Cell forms that are thin, flat, and tightly bound together usually have a ____________ function.

3. Which of the following statements about a hypothetical composite cell is (are) true?

a. It is necessary to construct a composite cell because cells vary so much, based on their function.

b. It contains structures that occur in many kinds of cells.

c. It contains structures that occur in all cells, although the characteristics of the structure may vary.

d. It is an actual cell type chosen because it occurs most commonly in the body.

4. The two major portions of the cell, each of which is surrounded by a membrane, are the_______________ and the_______________.

5. The organelles are located in the

a. nucleolus.

b. cytoplasm.

c. cell matrix.

d. cell membrane.

6. The mechanism that allows cells to communicate with other cells is _______________.

7. The cell membrane allows some substances to pass through it and excludes others. This is possible because the cell membrane is_______________.

8. The cell membrane is composed of a double layer of

a. protein molecules.

b. phospholipid molecules.

c. polysaccharide molecules.

d. amino acids.

9. The inner layer of the cell membrane composed of the fatty acid portion of lipid molecules is *impermeable* to molecules that are soluble in_______________.

10. Protein molecules embedded in the phospholipids of the cell membrane that span the cell membrane are known as_______________ or _______________ proteins.

11. Examples of ions that move across the cell membrane are _______________ and _______________.

12. The organelle that functions as a system of transport for materials from one part of the cytoplasm to another is the_______________.

13. Ribosomes function in the synthesis of protein molecules.

a. True

b. False

14. The Golgi apparatus is involved in the "packaging" of proteins for secretion to the (inside, outside) of the cell.

15. The mitochondria function in the release of _______________to the cells.

16. The enzymes of the lysosome

a. control cell reproduction.

b. digest bacteria and damaged cell parts.

c. release energy from where it is stored within the cell.

d. control the Krebs cycle.

17. Peroxisomes are abundant in the _______________ and the _______________.

18. Microfilaments are rods of protein involved in cellular _______________.

19. Which of the following statements about the centrosome is (are) true?

a. It is located near the nucleus.

b. The centrioles of the centrosome function solely in reproduction.

c. The centrosome is concerned with the distribution of chromosomes.

d. All of the above are true.

20. Cilia are found on the surface of

a. endothelial cells.

b. epithelial cells.

c. epidermal cells.

d. mucosa.

21. The structures suspended in the nucleoplasm of the nucleus are the _______________and the_______________.

22. The difference between active and passive mechanisms of movement through cell membranes is that active mechanisms require_______________.

23. The process that allows the movement of gases and ions from areas of higher concentration to areas of lower concentration until equilibrium has been achieved is called_______________.

24. The process by which lipid insoluble material moves through the cell membrane by using a carrier molecule is called_______________.

25. The process by which water moves across a semipermeable membrane from areas of low concentration of solute to areas of higher concentration is called________________________________.

26. A hypertonic solution is one that
a. contains a greater concentration of solute than the cell.
b. contains the same concentration of solute as the cell.
c. contains a lesser concentration of solute than the cell.

27. The process by which molecules are forced through a membrane by pressure that is greater on one side than on the other side is called________________________.

28. The process that uses energy to move ions across a concentration gradient from an area of lower concentration to an area of higher concentration is called____________________________.

29. The process by which cells engulf liquid molecules is called________________________.

30. A process that allows cells to take in molecules of solids is called___________________.

31. Once solid material is taken into a vacuole, which of the following statements best describes what happens?
a. A ribosome enters the vacuole and uses the amino acids in the "invader" to synthesize new proteins.
b. A lysosome combines with the vacuole and digests the enclosed solid material.
c. The vacuole remains separated from cytoplasm, and the solid material persists unchanged.
d. Oxygen enters the vacuole and burns the enclosed solid material.

32. The process that ensures duplication of DNA molecules during cell reproduction is___________________________.

33. Match these events with their correct descriptions.

____a. prophase
____b. metaphase
____c. anaphase
____d. telophase

1. Microtubules shorten; chromosomes are pulled toward centrioles.
2. Chromatin forms chromosomes; nuclear envelope and nucleolus break up and disperse.
3. Chromosomes elongate; nuclear membranes form around each chromosome set.
4. Chromosomes become arranged midway between centrioles; centromeres are broken.

34. The process by which cells develop unique characteristics in structure and function is called ________________________.

STUDY ACTIVITIES

Aids to Understanding Words

Define the following word parts. (p. 51)

cyt-

endo-

hyper-

hypo-

inter-

iso-

mit-

phag-

pino-

-som

3.1 Introduction (p. 51)

A. 1. The unit of life of human beings is the ________________________.

2. Explain your response.

B. In what ways do cells differ and why is this important?

3.2 Composite Cell (pp. 52–60)

A. Answer the following questions about the cell membrane. (pp. 52–54)

1. What are the three basic cell parts?

2. Chemically, the cell membrane is mainly composed of ____________ and ______________ with some ______________.

3. Describe the role of lipid molecules in making the cell membrane selectively permeable.

4. Where are carbohydrate molecules located?

5. What is their function?

B. Fill in the following table regarding the structure and function of organelles. (pp. 55–61)

Organelles	Structure	Function
Cell membrane		
Endoplasmic reticulum		
Ribosomes		
Golgi apparatus		
Mitochondria		
Lysosomes		
Peroxisomes		
Microfilaments and microtubules		
Centrosome		
Cilia and flagella		
Vesicles		
Nuclear envelope		
Nucleolus		
Chromatin		

3.3 Movements Through Cell Membranes (pp. 60–67)

A. Answer these questions concerning movement through cell membranes. (pp. 60–64)

1. When does diffusion stop?

2. What substances in the human body are transported by diffusion?

3. What is dialysis? Describe how it is used in an artificial kidney.

4. Describe facilitated diffusion.

5. How does osmosis differ from diffusion?

B. Below are three drawings of a red blood cell in solutions of varying tonicity. Label the tonicity of the solution in each drawing, and explain what is happening and why. (pp. 63–64)

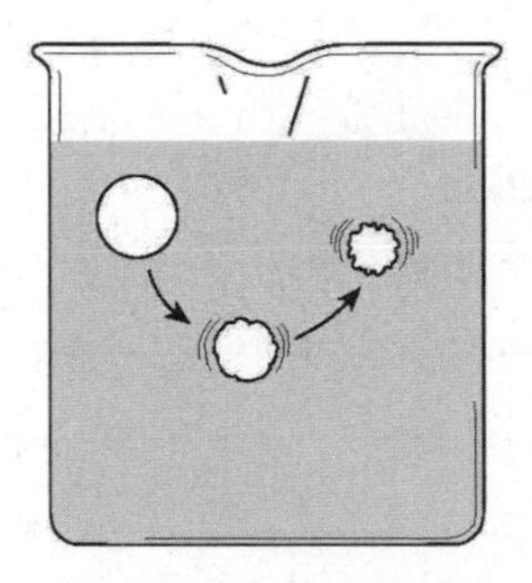

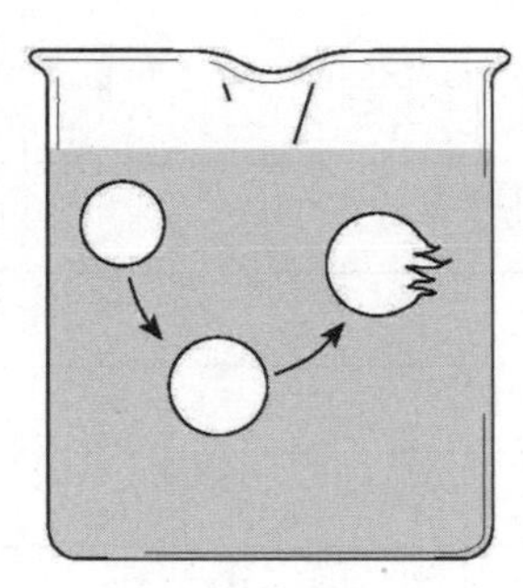

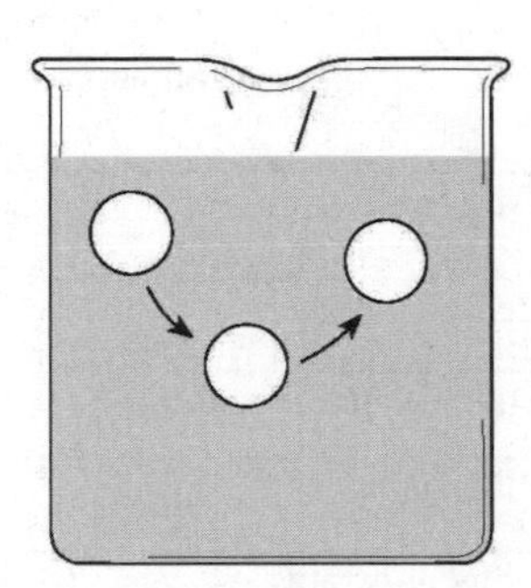

C. Answer these questions concerning the accompanying illustration. (p. 64)

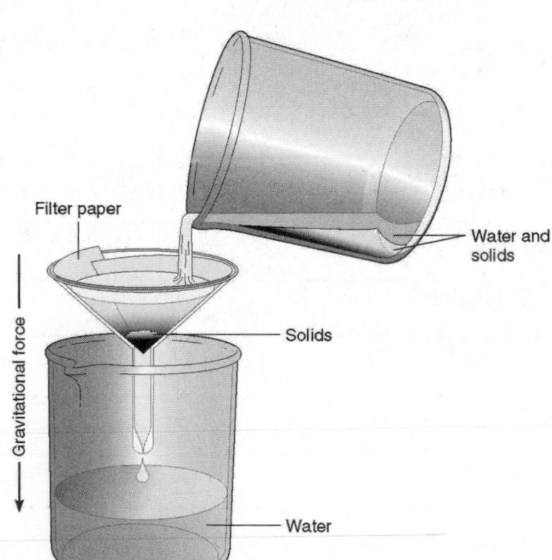

1. What process is illustrated here?

2. What provides the force needed to pull the liquid through the solids?

3. Where does this process occur within the body?

D. Answer the questions concerning the accompanying diagram. (pp. 65)

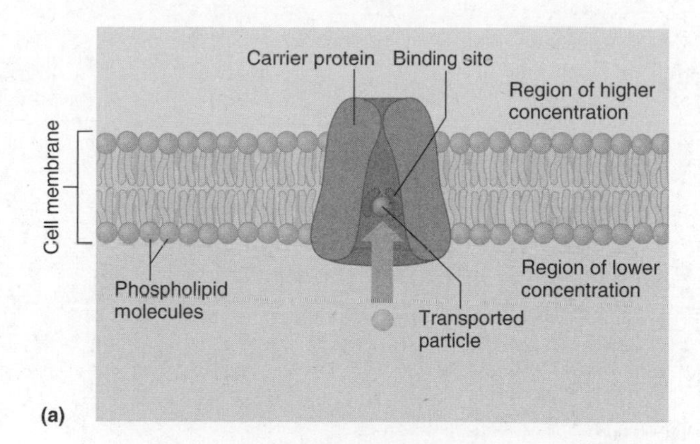

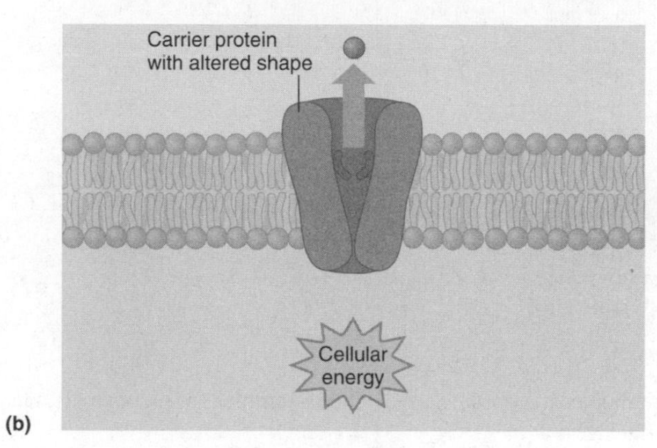

1. What transport mechanism is illustrated here?

2. What provides the necessary force for this process?

3. What is the source of this force?

4. How are molecules transported across the cell membrane?

5. What substances are transported by this mechanism?

E. Answer these questions concerning active transport. (p. 65)

1. In active transport, molecules move from regions of__________________ concentration to regions of __________________ concentration.

2. Compare active transport and facilitated diffusion.

F. Answer the questions concerning the accompanying diagram. (p. 66)

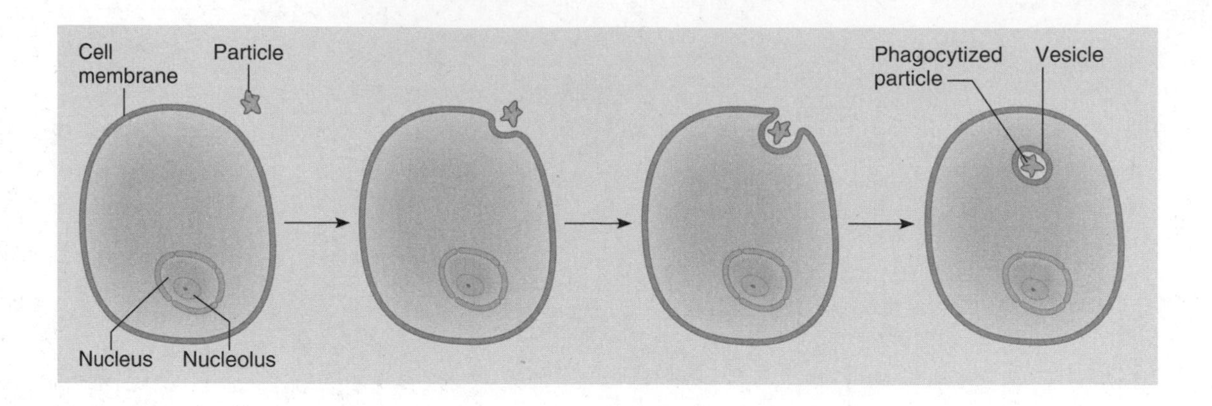

1. What mechanism is illustrated in this drawing?

2. What organelles are involved?

3. What kinds of foreign objects are transported? Give examples for pinocytosis, phagocytosis, and receptor-mediated endocytosis in your answer.

4. How is this mechanism important to cell survival?

3.4 The Cell Cycle (pp. 67–72)

A. 1. The series of changes a cell undergoes from its formation until its reproduction is called its________________________. (p. 67)

2. DNA is replicated during the ______________ phase of interphase.

B. Describe each of the following events in mitosis. (pp. 69–70)

1. Prophase, metaphase, anaphase, and telophase

2. What is the function of telomeres?

C. Answer the following questions regarding the cell cycle. (pp. 69–70.

1. Describe each of the following events in mitosis.

prophase

metaphase

anaphase

telophase

2. What is the function of telomeres.

3. Describe the events of cytoplasmic division.

D. Answer these questions concerning cell differentiation. (pp. 71–72)

1. The process by which cells develop differences in structure and function is known as__________________________________.

2. Compare a stem cell and a progenitor cell.

E. Describe the process of apoptosis. (p. 71)

Clinical Focus Question

How is knowledge of the mechanisms of cellular transport and cellular reproduction applied to the treatment of disease?

When you have finished the study activities to your satisfaction, retake the mastery test and compare your results with your initial attempt. If you are not satisfied with your performance, repeat the appropriate study activities.

CHAPTER 4
CELLULAR METABOLISM

OVERVIEW

This chapter explains the processes of cellular metabolism. It explains the role of genes in these processes, the function of enzymes, and enzymatic control of reactions and describes the concept of the active site (learning outcomes 1–3). Energy is essential for these processes. The events of cellular respiration are discussed as are the structure and function of ATP and the role of oxygen (learning outcomes 4 and 5). The metabolic pathways of foods are described (learning outcome 6). Finally, the role of genetic material, including DNA and RNA, in cellular metabolism and the replication of these essential genetic materials end the discussion of these metabolic processes (learning outcomes 7–9).

LEARNING OUTCOMES

After you have studied this chapter, you should be able to

4.1 Introduction
1. Briefly explain the function of metabolism.

4.2 Metabolic Reactions
2. Compare and contrast anabolism and catabolism.

4.3 Control of Metabolic Reactions
3. Describe how enzymes control metabolic reactions.

4.4 Energy for Metabolic Reactions
4. Explain how cellular respiration releases chemical energy.
5. Describe how energy in the form of ATP becomes available for cellular activities.

4.5 Metabolic Pathways
6. Describe the general metabolic pathways of carbohydrates, lipids, and proteins.

4.6 DNA (Deoxyribonucleic Acid)
7. Describe how DNA molecules store genetic information.
8. Describe how DNA molecules are replicated.

4.7 Protein Synthesis
9. Describe the steps of protein synthesis.

FOCUS QUESTION

How do cells carry on life processes?

MASTERY TEST

Now take the mastery test. Do not guess. As soon as you complete the test, correct it. Note your successes and failures so that you can read the chapter to meet your learning needs.

1. The proteins that control the reactions of metabolism are
   a. amino acids.
   b. catalysts.
   c. enzymes.
   d. substrates.
2. The metabolic process that synthesizes materials needed for cellular growth is called_______________________________.
3. The metabolic process that breaks down complex molecules into simpler ones is called______________________________.
4. The process by which two molecules are joined together to form a more complex molecule is called
   a. dehydration synthesis.
   b. chemical bonding.
   c. atomization.
5. Glycerol and fatty acids become bonded to form water and
   a. cholesterol.
   b. lard.
   c. fat molecules.
   d. wax.
6. Amino acid molecules are joined by a peptide bond to form water and_________________________.
7. The process by which water is added to a complex molecule to break it down, as represented by the equation $C_{12}H_{22}O_{11} + H_2O = C_6H_{12}O_6 + C_6H_{12}O_6$, is called_____________________________________.
8. A substance that decreases the amount of energy necessary to begin a chemical reaction is a(n) _______________________.
9. Enzymes are composed of
   a. lipids.
   b. carbohydrates.
   c. proteins.
   d. inorganic salts.
10. An enzyme acts only on a particular substance that is called a
    a. binding site.
    b. substrate.
    c. complement.
    d. histologue.
11. An enzyme's ability to recognize the substance upon which it will act seems to be based on
    a. atomic weight.
    b. molecular shape.
    c. structural formula.
12. The suffix that identifies a substance as an enzyme is ______________.
13. The ability of an enzyme to recognize its specific substrate is dependent upon its __________________ or __________________.
14. The speed of an enzyme-controlled reaction depends upon the number of ______________ and _______________molecules present.
15. A substance needed to convert an inactive form of an enzyme to an active form is called a(n) ____________________.
16. The form of energy utilized by most cellular processes is
    a. chemical.
    b. electrical.
    c. thermal.
    d. mechanical.
17. The process by which energy is released in the cell is called_________________________.
18. The initial phase of respiration that occurs in the cytosol and produces two 3-carbon pyruvic acid molecules plus energy is called___________________________ or _____________________________.
19. The second phase of respiration that occurs in the mitochondria and produces carbon dioxide, water, and energy is called_____________________________.
20. What element is needed for this second phase of respiration to take place?
21. Name the storage place for the energy released by cellular respiration.
22. A particular sequence of enzyme-controlled reactions is called a(n)_________________________________.
23. We inherit traits from our parents because
    a. DNA contains genes that are the carriers of inheritance.
    b. genes tell the cells to construct protein in a unique way for each individual.
    c. our species, *Homo sapiens,* reproduces sexually.
24. A molecule consisting of a double spiral with sugar and phosphates forming the outer strands and organic bases joining the two strands is____________________.

25. List the four organic bases of DNA nucleotides.

26. DNA molecules are located in the_________________. Protein synthesis takes place in the_____________________.

27. Two types of RNA are ________________RNA and ________________RNA.

28. In a molecule of RNA, the thymine nucleotide of DNA is replaced by____________________.

29. The function of RNA is to
a. form lipids, such as cholesterol.
b. guide the breakdown of polysaccharides.
c. control the bonding of amino acids.

30. When the genetic material of a cell is altered, the result may be a(n)_____________________ .

31. The rate at which a metabolic pathway functions is determined by a(n) ___________________that is present in limited quantity.

32. Proteomics is the study of
a. protein synthesis.
b. gene reproduction.
c. protein conservation.
d. gene expression.

33. DNA replication takes place during the ____________ of the cell cycle.

STUDY ACTIVITIES

Aids to Understanding Words

Define the following word parts. (p. 77)

an-

ana-

cata-

mut-

-zym

4.1 Introduction (p. 77)

A. Life is maintained by a series of _______________ reactions.

B. The reactions are controlled by proteins called _______________.

4.2 Metabolic Reactions (pp. 77–78)

A. Answer these questions concerning anabolic metabolism. (pp. 77–78)

1. Define *anabolic metabolism.*

2. The following formula is an example of anabolic metabolism. Label the formula, and identify the process illustrated.

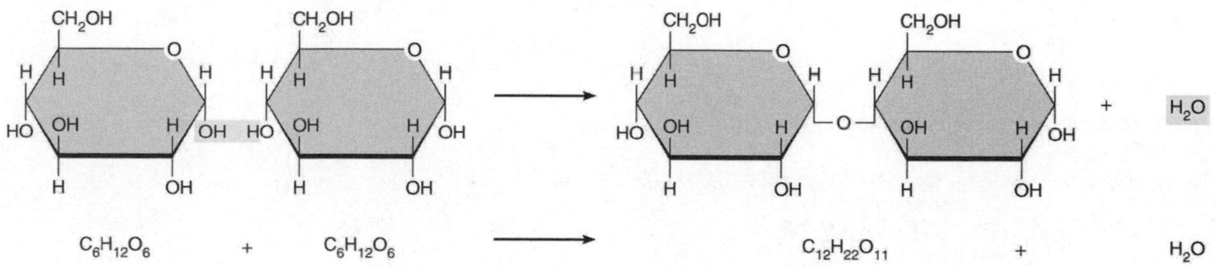

$$C_6H_{12}O_6 + C_6H_{12}O_6 \rightarrow C_{12}H_{22}O_{11} + H_2O$$

B. Answer these questions concerning catabolism. (p. 78)

1. Define *catabolism.*

2. What process is illustrated here?

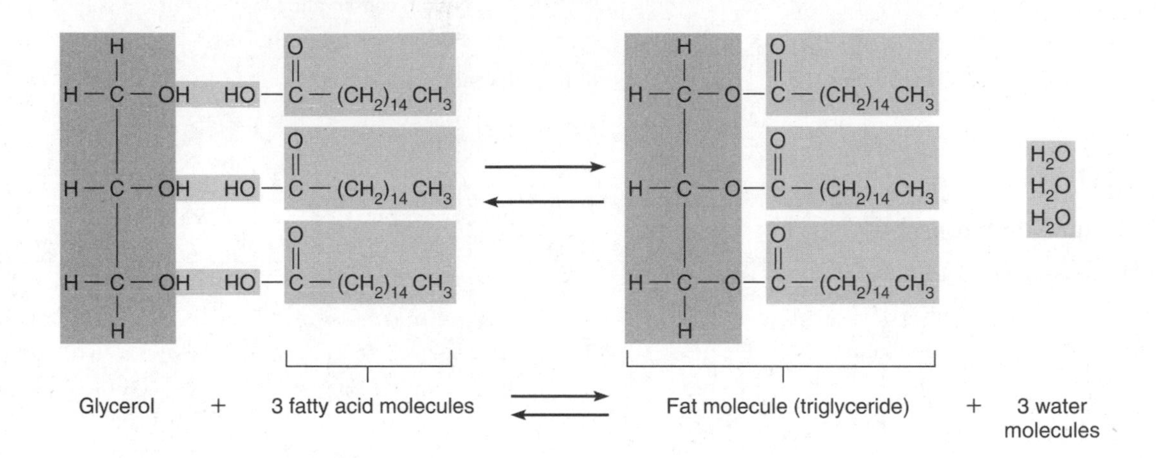

4.3 Control of Metabolic Reactions (pp. 79–80)

A. Enzymes promote chemical reactions in cells by ______________________ the amount of ______________________needed to initiate a reaction. (p. 79)

B. What is the relationship between an enzyme and a substrate? Explain how this relationship works. (p. 79)

C. What is a coenzyme? What substances are coenzymes? (p. 80)

D. List the agents that can denature enzymes. (p. 80)

4.4 Energy for Metabolic Reactions (pp. 80–82)

A. Answer these questions concerning energy and its release. (p. 80)

1. What is energy?

2. List six common forms of energy.

3. The form of energy used by cell processes is________________________________.

4. This energy is released by the process of ________________________________.

B. Fill in the following table, comparing aerobic and anaerobic respiration. (p. 82)

	Anaerobic	**Aerobic**
Location of reaction within cell		
How released energy is captured		
Number of molecules formed		

4.5 Metabolic Pathways (pp. 82–83)

A. What is a metabolic pathway? (p. 82)

B. How are metabolic pathways regulated? (p. 82)

4.6 DNA (Deoxyribonucleic Acid) (pp. 83–85)

A. What is a gene? (p. 84)

B. How are genes necessary to cell metabolism? (pp. 83–84)

C. What is the genome? (p. 84)

D. What is the structure of DNA? (p. 84)

E. What are the possible bases of nucleotides in DNA? (p. 84)

F. Draw the matching strand of DNA for the strand illustrated here.

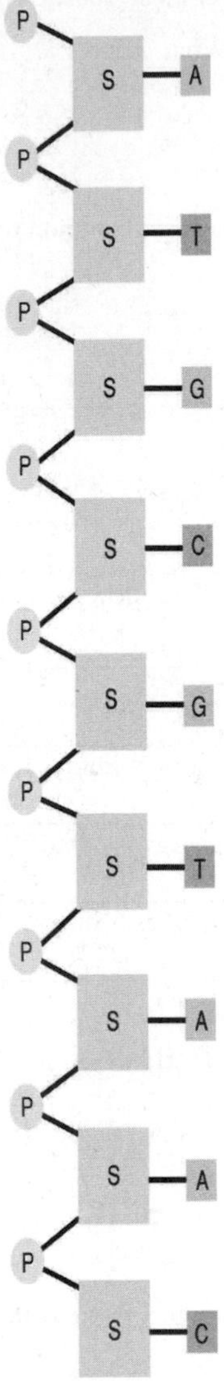

G. DNA molecules are replicated during ____________________ of the cell cycle. (p. 84)

H. Explain the process of DNA replication. (pp. 84–85)

4.7 Protein Synthesis (pp. 85–89)

A. Where does protein synthesis occur? (p. 85)

B. How does RNA differ from DNA? (p. 86)

C. Describe the roles of messenger RNA and transfer RNA in protein synthesis. (pp. 86–88)

D. What is the role of the proteasome and how is this important for normal metabolism? (p. 88)

Clinical Focus Question

How do an individual's health history, past medical history, and medication history assist a health-care professional in planning an individual's health care?

When you have finished the study activities to your satisfaction, retake the mastery test and compare your results with your initial attempt. If you are not satisfied with your performance, repeat the appropriate study activities.

CHAPTER 5
TISSUES

OVERVIEW

This chapter deals with the simplest level of organization of cells—tissues. It explains the types of tissue that occur in the human body, the general functions of each of these types of tissue, and the organs in which the various types of tissue and membranes occur (learning outcomes 1–9).

The characteristics of a tissue remain the same regardless of where it occurs in the body. Knowledge of these characteristics is basic to understanding how a specific tissue contributes to the function of an organ.

LEARNING OUTCOMES

After you have studied this chapter, you should be able to

5.1 Introduction

1. List the four major tissue types and tell where each is located in the body.

5.2 Epithelial Tissues

2. Describe the general characteristics and functions of epithelial tissue.
3. Name the types of epithelium, and for each type, identify an organ in which that type is found.
4. Explain how glands are classified.

5.3 Connective Tissues

5. Compare and contrast the general components, cells, fibers, and extracellular matrix (where applicable) in each type of connective tissue.
6. Describe the major functions of each type of connective tissue.

5.4 Types of Membranes

7. Distinguish among the four major types of membranes.

5.5 Muscle Tissues

8. Distinguish among the three types of muscle tissues.

5.6 Nervous Tissues

9. Describe the general characteristics and functions of nervous tissues.

FOCUS QUESTION

How is tissue related to the organization of the body?

MASTERY TEST

Now take the mastery test. Do not guess. As soon as you complete the test, correct it. Note your successes and failures so that you can read the chapter to meet your learning needs.

1. List the four major types of tissue found in the human body.
2. Cells in a tissue are (similar, dissimilar).
3. The function of epithelial tissue is to
   a. support body parts.
   b. cover body surfaces.
   c. bind body parts together.
   d. form the framework of organs.

4. Which of the following statements about epithelial tissue is (are) true?

a. Epithelial tissue has no blood vessels.

b. Epithelial cells reproduce slowly.

c. Epithelial cells are nourished by substances diffusing from connective tissue.

d. Injuries to epithelial tissue heal rapidly as new cells replace damaged cells.

5. The underside of epithelial tissue is formed of nonliving tissue called the ________________________.

6. Match the following types of epithelial cells with their correct location.

____a. simple squamous epithelium
____b. simple cuboidal epithelium
____c. simple columnar epithelium
____d. pseudostratified columnar epithelium
____e. stratified squamous epithelium

1. lining of the ducts of salivary glands
2. lining of respiratory passages
3. epidermis of the skin
4. air sacs of lungs, walls of capillaries
5. lining of digestive tract

7. The inner lining of the urinary bladder and the passageways of the urinary tract are composed of ____________________.

8. A gland that secretes its products into ducts opening into an external or internal surface is called a(n)____________________gland.

9. A merocrine gland secretes its product by ________________.

10. The functions of connective tissue is (are)

a. support.

b. protection.

c. to fill spaces.

d. fat storage.

11. Fibroblasts and mast cells found in connective tissue are (fixed, wandering) cells.

12. The connective tissue cells that produce fibers are

a. mast cells.

b. macrophages.

c. fibroblasts.

13. The major structural protein of the body and of white connective tissue is__________________.

14. Yellow connective tissue that can be stretched and returned to its original shape is__________________.

15. The most common cells of loose connective tissue are __________________________.

16. Which of the following statements is (are) true regarding adipose tissue?

a. It is a specialized form of loose connective tissue.

b. It occurs around the kidneys, behind the eyeballs, and around various joints.

c. It serves as a conserver of body heat.

d. It serves as a storehouse of energy for the body.

17. The cartilage found in the tip of the nose is ______________________ cartilage.

18. The type of cartilage in the intervertebral discs is __________________.

19. Because of the nature of its blood supply, injured cartilage heals (quickly, slowly).

20. The most rigid connective tissue is ______________________________.

21. The intercellular material of vascular tissue is __________________________.

22. The three types of muscle tissue are _______________, ________________, and ________________.

23. Coordination and regulation of body functions are the functions of ______________ tissue.

24. List the four major types of membranes.

25. Serous membranes are located

a. around the structures of the dorsal cavity.

b. in body cavities that are completely closed to the outside of the body.

c. wherever two bones come together.

26. Mucous membranes are located

a. around the organs of the respiratory system.

b. in the lining of cavities and tubes that have openings to the outside of the body.

c. in areas where two surfaces meet.

STUDY ACTIVITIES

Aids to Understanding Words

Define the following word parts. (p. 95)

adip-

chondr-

-cyt

epi-

-glia

inter-

macr-

os-

pseud-

squam-

strat-

5.1 Introduction (p. 95)

A. What are tissues?

B. List the major tissue types and their functions.

5.2 Epithelial Tissues (pp. 95–102)

A. List four functions of epithelial tissue. (p. 95)

B. What is the function of the basement membrane? (p. 95)

C. Answer these questions concerning simple squamous epithelium. (p. 96)

1. Describe the structure of simple squamous epithelium.

2. Where is simple squamous epithelium found? What are the functions of this tissue?

D. Answer these questions concerning simple cuboidal epithelium. (p. 96)

1. Describe the structure of simple cuboidal epithelium.

2. Where is this type of tissue found? What are the functions of this tissue?

E. Answer these questions concerning simple columnar epithelium. (p. 97)

1. Describe the structure of simple columnar epithelium.

2. Where is this tissue located? What is the function of simple columnar epithelium?

F. Answer these questions concerning pseudostratified columnar epithelium. (p. 98)

1. Microscopic, hairlike projections called______________________are a characteristic of columnar epithelium.

2. Where is this tissue found?

G. Answer these questions concerning stratified squamous epithelium. (p. 98)

1. Describe the structure of stratified squamous epithelium.

2. Where is this tissue found?

H. What is the special characteristic of transitional epithelium? (p. 98)

I. Describe glandular epithelium. (pp. 101–102)

J. Fill in the following table regarding the different types of glands. (p. 101)

Type of gland	Description of secretion	Examples
Merocrine		
Apocrine		
Holocrine		

5.3 Connective Tissues (pp. 102–109)

A. What are the functions and structure of connective tissue? (p. 102)

B. What are the functions of fibroblasts, macrophages, and mast cells? (pp. 102–103)

C. How do collagenous fibers and elastic fibers differ? (p. 103)

D. What is the difference between a ligament and a tendon? (p. 103)

E. Where is adipose connective tissue found, and what is its function? (pp. 104–106)

F. Fill in the following table regarding the different types of cartilage. (pp. 107–108)

Type	Location	Function
Hyaline		
Elastic		
Fibrocartilage		

G. Answer these questions concerning bone. (p. 108)

1. What are the characteristics of bone?

2. Bone injuries heal relatively rapidly. Why is this true?

H. Answer these questions concerning blood (vascular connective tissue). (p. 108)

1. What is the intercellular material of vascular connective tissue?

2. What cells are found in the intercellular material?

3. Describe the functions of the connective tissue matrix.

5.4 Types of Membranes (p. 110)

A. List the three types of epithelial membranes. (p. 110)

B. Where is each membrane found? (p. 110)

5.5 Muscle Tissues (pp. 110–111)

A. What are the characteristics of muscle tissue? (p. 110)

B. Fill in the following table regarding the different types of muscle. (pp. 110–111)

Type of muscle	Structure	Control	Location
Skeletal			
Cardiac			
Smooth			

5.6 Nervous Tissues (pp. 111–112)

A. What is the basic cell of nervous tissue called? (p. 111)

B. What is the function of neuroglial cells in nervous tissue? (p. 112)

C. What is the function of nervous tissue? (pp. 111–112)

Clinical Focus Question

Kenny, age ten, is tall for his age, and his large muscles are well developed. He is very anxious to play football with a group of boys who are fourteen and fifteen. What problems may result for Kenny? What tissues are most likely to be at risk for injury?

When you have finished the study activities to your satisfaction, retake the mastery test and compare your results with your initial attempt. If you are not satisfied with your performance, repeat the appropriate study activities.

CHAPTER 6

SKIN AND THE INTEGUMENTARY SYSTEM

OVERVIEW

This chapter describes the skin and its accessory organs. It begins by defining the nature of an organ and associates that concept with the skin as an organ of the integumentary system (learning outcome 1). It continues by explaining the structure and function of the various layers of the skin and summarizes the factors that determine skin color (learning outcomes 2–4). The accessory structures of the skin and their function, the regulation of body temperature, and wound healing are the final topics in this chapter (learning outcomes 4–7).

LEARNING OUTCOMES

After you have studied this chapter, you should be able to

6.1 Introduction

1. Describe what constitutes an organ and name the large organ of the integumentary system.

6.2 Skin and Its Tissues

2. List the general functions of the skin.
3. Describe the structure of the layers of the skin.
4. Summarize the factors that determine skin color.

6.3 Accessory Structures of the Skin

5. Describe the anatomy and physiology of each accessory structure of the skin.

6.4 Regulation of Body Temperature

6. Explain how the skin helps regulate body temperature.

6.5 Healing of Wounds

7. Describe wound healing.

FOCUS QUESTION

How do the structure and function of the skin earn it the title of *the body's first line of defense?*

MASTERY TEST

Now take the mastery test. Do not guess. As soon as you complete the test, correct it. Note your successes and failures so that you can read the chapter to meet your learning needs.

1. The outer layer of skin is called the ______________________________.
2. The inner layer of skin is called the ______________________________.
3. The masses of connective tissue beneath the inner layers of skin are called the ______________________________.
4. The outermost layer of the epidermis is the
   a. keratin.
   b. stratum corneum.
   c. stratum basale.
   d. epidermis.
5. The pigment that helps protect the deeper layers of the epidermis is
   a. melanin.
   b. trichosiderin.
   c. biliverdin.
   d. bilirubin.

6. People with light complexions have (greater, fewer, equal) numbers of melanocytes when compared to people with dark complexions.

7. Blood vessels supplying the skin are located in the ________________________.

8. Smooth muscle cells that stand hairs on end in response to cold are known as________________________________.

9. The glands usually associated with hair follicles are
   a. apocrine glands.
   b. endocrine glands.
   c. sebaceous glands.
   d. ceruminous glands.

10. Nails are produced by epidermal cells that undergo ___________________________.

11. Where are the eccrine sweat glands most commonly found?
    a. the forehead
    b. the groin and the axilla
    c. the neck and back
    d. evenly distributed over the body surface

12. The mammary glands of the breast that produce milk are modified ________________ glands.

13. An irregularly shaped lesion with variegated color that develops on sun-exposed areas is a(n) ___________________.

14. The sweat glands associated with regulation of body temperature are the
    a. endocrine glands.
    b. eccrine glands.
    c. sebaceous glands.
    d. apocrine glands.

15. Which of the following organs produce the most heat?
    a. kidneys
    b. bones
    c. muscles
    d. lungs

16. Sponging the skin with water helps increase the loss of body heat by
    a. evaporation.
    b. convection.
    c. conduction.
    d. radiation

17. What are the signs of inflammation?

18. Fibroblasts must migrate into a (shallow, deep) cut to heal the skin defect.

STUDY ACTIVITIES

Aids to Understanding Words

Define the following word parts. (p. 117)

cut-

derm-

epi-

follic-

kerat-

melan-

seb-

6.1 Introduction (p. 117)

A. What is an organ?

B. What are the components of the integumentary system?

6.2 Skin and Its Tissues (pp. 117–122)

A. List the functions of the skin. (p. 117)

B. What kinds of tissue are found in the skin? (p. 117)

C. Label the layers of skin in the illustration below. (p. 119)

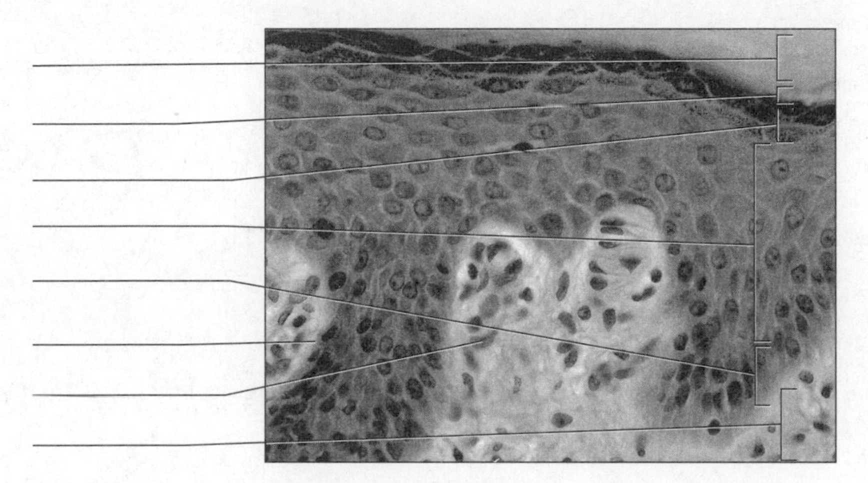

D. Answer the following questions about the layers of the epidermis. (pp. 119–120)

1. What is the function of the stratum basale?

2. What is the function of the stratum corneum?

3. A callus or a corn is the result of a(n) ________________________ in cell reproduction in response to ____________________________ or ____________________________.

E. Deep layers of skin are protected from the ultraviolet portion of sunlight by ____________________. (p. 119)

F. Answer these questions concerning factors that influence skin color. (pp. 120–122)

1. What environmental factors influence skin color?

2. What physiological factors influence skin color?

G. Describe the structure and function of the dermis. (p. 122)

H. Describe cutaneous carcinomas and cutaneous melanomas. (p. 121)

6.3 Accessory Structures of the Skin (pp. 122–124)

A. Where is the growing portion of the nail located? (p. 122)

B. How is hair formed in the hair follicle? (pp. 122–123)

C. Describe how hair responds to cold temperature or strong emotion. (p. 123)

D. Where are sebaceous glands located, and what is the function of the substance they secrete? (p. 124)

E. Label the following parts of a hair follicle on the illustration below (p. 123): hair shaft, hair follicle, arrector pili muscle, sebaceous gland, dermal blood vessels, hair root, region of cell division, eccrine sweet gland, basement membrane.

F. Compare apocrine and eccrine sweat glands in relation to location, association with other skin structures, and activating stimuli. (p. 124)

6.4 Regulation of Body Temperature (p. 125)

Describe the roles of the nervous, muscular, circulatory, and respiratory systems in heat regulation.

6.5 Healing of Wounds (pp. 125–127)

A. Describe the process of inflammation. What function does this process serve? (p. 125)

B. Compare the healing of a shallow wound and of a wound that extends into the dermis. (pp. 125–126)

Clinical Focus Questions

A. The ability to maintain body temperature is dependent on the integumentary system. Compare the integumentary systems of neonates and the elderly and identify the mechanisms that place these two age groups at risk for body temperature disturbances.

B. Based on your current knowledge, list factors that you think might regulate body temperature.

When you have finished the study activities to your satisfaction, retake the mastery test and compare your results with your initial attempt. If you are not satisfied with your performance, repeat the appropriate study activities.

UNIT 2

SUPPORT AND MOVEMENT

CHAPTER 7

SKELETAL SYSTEM

OVERVIEW

This chapter deals with the skeletal system—the bones that form the framework for the body. It explains the function and structure of bones (learning outcomes 1, 2, and 4). The development of different types of bone is also explained (learning outcome 3). This chapter describes skeletal organization and the location of specific bones within various parts of the skeleton (learning outcomes 5 and 6). Various types of joints and the movements made possible by these joints (learning outcomes 7–9) are also described in this chapter.

Movement is a characteristic of living things. A study of the skeletal system is necessary to understand how complex organisms, such as humans, are organized to accomplish movement.

LEARNING OUTCOMES

After you have studied this chapter, you should be able to

7.1 Introduction

1. List the active tissues in a bone.

7.2 Bone Structure

2. Describe the macroscopic and microscopic structure of a long bone, and list the functions of these parts.

7.3 Bone Development and Growth

3. Distinguish between intramembranous and endochondral bones, and explain how such bones develop and grow.

7.4 Bone Function

4. Discuss the major functions of bones.

7.5 Skeletal Organization

5. Distinguish between the axial and appendicular skeletons and name the major parts of each.

7.6–7.12 Skull–Lower Limb

6. Locate and identify the bones and the major features of the bones that compose the skull, vertebral column, thoracic cage, pectoral girdle, upper limb, pelvic girdle, and lower limb.

7.13 Joints

7. Classify joints according to the type of tissue binding the bones together, describe the different joint characteristics, and name an example of each joint type.
8. List six types of synovial joints, and describe the actions of each.
9. Explain how skeletal muscles produce movements at joints, and identify several types of joint movements.

FOCUS QUESTION

How do your bones and joints help you to get out of bed and to your anatomy and physiology class?

MASTERY TEST

Now take the mastery test. Do not guess. As soon as you complete the test, correct it. Note your successes and failures so that you can read the chapter to meet your learning needs.

1. Which of the following is *not* an active tissue found in bone?
   a. cartilage
   b. cuboidal epithelium
   c. blood
   d. nervous tissue

2. The shaft of a long bone is the
   a. epiphysis.
   b. diaphysis.

3. To what part of the bone do tendons and ligaments attach?
   a. bursae
   b. epiphysis
   c. cartilage
   d. periosteum

4. Bone that consists mainly of tightly packed tissue is called ______________________bone.

5. Bone that consists of numerous branching bony plates separated by irregular spaces is called _______________bone.

6. The medullary cavity of a long bone is filled with ______________________________.

7. Bones that develop from layers of membranous connective tissue are called ______________________________.

8. An example of a sesamoid bone is the ________________.

9. Bones that develop from masses of hyaline cartilage are called _________________________________.

10. The band of cartilage between the primary and secondary ossification centers in long bones is called the
    a. osteoblastic band.
    b. calcium disc.
    c. periosteal plate.
    d. epiphyseal plate.

11. The cells that form new bone are called ___________________; the cells that break down bone are called__________________.

12. Once bone formation is complete, the bone (remains stable, is remodeled) throughout life.

13. When a bone is fractured, a hematoma is formed from blood escaping from
    a. the periosteum.
    b. bone marrow.
    c. blood vessels within the bone.
    d. surrounding soft tissue.

14. List the major factors that influence bone growth and development.

15. The gap between broken ends of a fractured bone is filled by a _________________________.

16. To accomplish movement, bones and muscles function together to act as __________________________________.

17. Which of the following bones contain(s) red marrow for blood cell formation in a healthy adult?
    a. pelvis
    b. small bones of the wrist
    c. ribs
    d. shaft of long bones

18. Which of the following substances is *not* normally found in bone?
    a. potassium
    b. calcium
    c. lead
    d. magnesium

19. Calcium is important in
    a. muscle contration.
    b. regulation of thyroid function.
    c. nerve impulse conduction.
    d. blood cell formation.

20. List the major parts of the axial skeleton.

21. List the major parts of the appendicular skeleton.

22. The part of the spinal column in which the vertebrae are fused is the
    a. cervical spine.
    b. thoracic spine.
    c. sacrum.
    d. lumbar spine.

23. The only movable bone of the skull is the
   a. nasal bone.
   b. mandible.
   c. maxilla.
   d. vomer.

24. The bone that forms the back of the skull and joins the skull along the lambdoid suture is the ________________________ bone.

25. The upper jaw is formed by the ________________________________ bones.

26. The membranous areas (soft spots) of an infant's skull are called ____________________________.

27. What parts of the vertebral column act(s) as (a) shock absorbers?
   a. vertebral bodies
   b. intervertebral discs
   c. lamina
   d. spinous processes

28. Which of the vertebrae support the most weight?
   a. cervical
   b. thoracic
   c. lumbar
   d. sacral

29. The functions of the thoracic cage include
   a. production of blood cells.
   b. contribution to breathing.
   c. protection of heart and lungs.
   d. support of the shoulder girdle.

30. True ribs articulate with the ____________________ and ______________________ on a vertebrae and the____________________.

31. The pectoral girdle is made of two ________________________ and two ______________________.

32. The _______________________________ crosses over the ulna when the palm of the hand faces backward.

33. The wrist consists of
   a. eight carpal bones.
   b. five metacarpal bones.
   c. fourteen phalanges.
   d. distal segments of the radius and the ulna.

34. When the hands are placed on the hips, they are placed over the
   a. iliac crest.
   b. acetabulum.
   c. ischial tuberosity.
   d. ischial spines.

35. The longest bone in the body is the
   a. tibia.
   b. fibula.
   c. femur.
   d. patella.

36. The lower end of the fibula can be felt as an ankle bone. The correct name for this bony feature is the
   a. head of the fibula.
   b. lateral malleolus.
   c. talus.
   d. lesser trochanter.

37. A synovial membrane is found in
   a. immovable joints.
   b. slightly movable joints.
   c. freely movable joints.

38. The function of bursae is to
   a. act as shock absorbers.
   b. facilitate movement of tendons over bones.
   c. reduce friction between bony surfaces.
   d. protect joints from infection.

39. The type of joint that permits the widest range of motion is
   a. ball-and-socket.
   b. gliding.
   c. condyloid.
   d. pivot.

40. Moving the parts at a joint so that the angle between them is increased is called
   a. flexion.
   b. extension.
   c. elevation.
   d. abduction.

STUDY ACTIVITIES

Aids to Understanding Words

Define the following word parts. (p. 133)

acetabul-

ax-

-blast

carp-

-clast

condyl-

corac-

cribr-

crist-

fov-

glen-

inter-

intra-

meat-

odont-

poie-

7.1 Introduction (p. 133)

List the living tissues of bone.

7.2 Bone Structure (pp. 133–135)

A. Label the following parts in the accompanying drawing of a long bone (p. 134): diaphysis, articular cartilage, spongy bone, compact bone, medullary cavity, yellow marrow, periosteum, epiphyseal discs, proximal epiphysis, distal epiphysis, space occupied by red marrow.

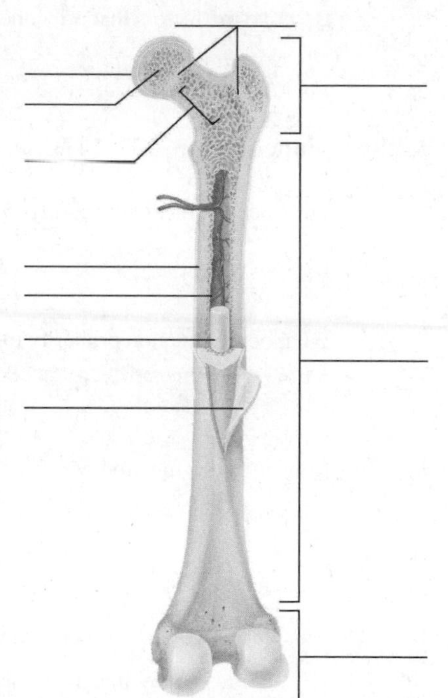

B. Describe the classification of bones by their shape. (p. 133)

C. How does the structure of bone make its function possible?

D. The vascular fibrous tissue covering the bone that functions in the formation and repair of bone tissue is called the ______________________________. (p. 133)

E. What is the structural difference between compact and spongy bone? (pp. 133–134)

F. Describe the microscopic structure of bone tissue (pp. 134–135)

7.3 Bone Development and Growth (pp. 135–137)

A. What bones are intramembranous bones? How do these develop? (pp. 135–136)

B. What bones are endochondral bones? How do these develop? *Be sure to include descriptions of the primary ossification center, the secondary ossification center, and the epiphyseal disc.* (pp. 136–137)

C. Answer these questions concerning ossification. (p. 137)

1. When is ossification complete?

2. Describe the homeostasis of bone tissue.

D. How is the fracture of a bone healed? (pp. 136–137)

E. Describe the factors that affect bone development and growth. (p. 137)

7.4 Bone Function (pp. 137–141)

A. What bones function primarily to provide support? (p. 137)

B. What bones function primarily to protect viscera? (p. 137)

C. How do bones function with muscles to produce movement? (p. 140)

D. Answer these questions concerning blood cell formation. (pp. 140–141)

1. Where are blood cells formed in the embryo? In the infant? In the adult?

2. What is the difference between red and yellow marrow?

E. Answer these questions concerning the inorganic compounds in bone. (p. 141)

1. What are the major inorganic salts stored in bone? What other salts and heavy metals can also be stored in bone?

2. How is calcium released from bone so that it is available for physiological processes?

7.5 Skeletal Organization (pp. 142–144)

A. What are the two major divisions of the skeleton? (p. 142)

B. List the bones found in each of these major divisions. (pp. 142–143)

7.6 Skull (pp. 144–149)

A. Answer these questions concerning the number of bones in the skull. (p. 144)

1. How many bones are found in the human skull?

2. How many of these bones are found in the cranium?

3. How many are found in the facial skeleton?

B. Answer these questions concerning the cranial bones. (pp. 144–147)

1. Using your own head or that of a partner, locate the following cranial bones and identify the suture lines that form their boundaries: occipital bone, temporal bones, frontal bones, parietal bones.

2. What are the remaining two bones of the cranium? Where are they located?

C. Answer these questions concerning the facial bones. (pp. 148–149)

1. Using your own head or that of a partner, locate the following facial bones: maxilla, palatine, zygomatic, lacrimal bones, nasal bones, vomer, inferior nasal conchae, mandible.

2. Which of the facial bones is the only movable bone of the skull?

3. Describe the differences between the infant and the adult skull.

7.7 Vertebral Column (pp. 149–153)

A. What is the function of the vertebral column? What is the function of intervertebral discs? (p. 149)

B. Label the following parts of the accompanying diagram (p. 152): lamina, transverse process, spinous process pedicle, superior articulating process, body, vertebral foramen.

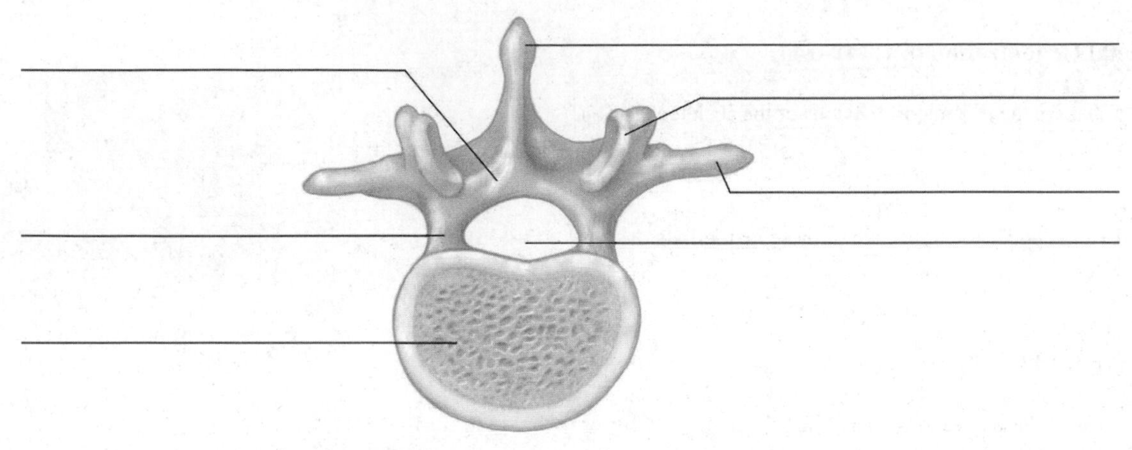

Lumbar vertebra

C. List the features of the thoracic vertebrae that are unique and explain why they are necessary. (pp. 151–152)

D. List the features of the lumbar vertebrae that are unique and explain why they are necessary. (p. 152)

E. Locate and describe the sacrum and the coccyx. (pp. 152–153)

7.8 Thoracic Cage (pp. 153–155)

A. Name the bones of the thoracic cage. (p. 153)

B. Describe the differences between true and false ribs. (pp. 153–154)

C. Describe the sternum, including the manubrium, body, and xiphoid process. Locate these structures on yourself. (p. 155)

7.9 Pectoral Girdle (p. 155)

Using yourself or a partner, locate and list the bones of the pectoral girdle. What is the function of the pectoral girdle? (p. 155)

7.10 Upper Limb (pp. 155–158)

A. Using yourself or a partner, locate, list, and describe the bones of the upper limb. (p. 155)

B. Label the following parts in the drawing below (p. 159): phalanges, metacarpals, carpals, pisiform, triquetrum, hamate, lunate, capitate, scaphoid, trapezoid, trapezium, radius, ulna, proximal phalanx, middle phalanx, distal phalanx.

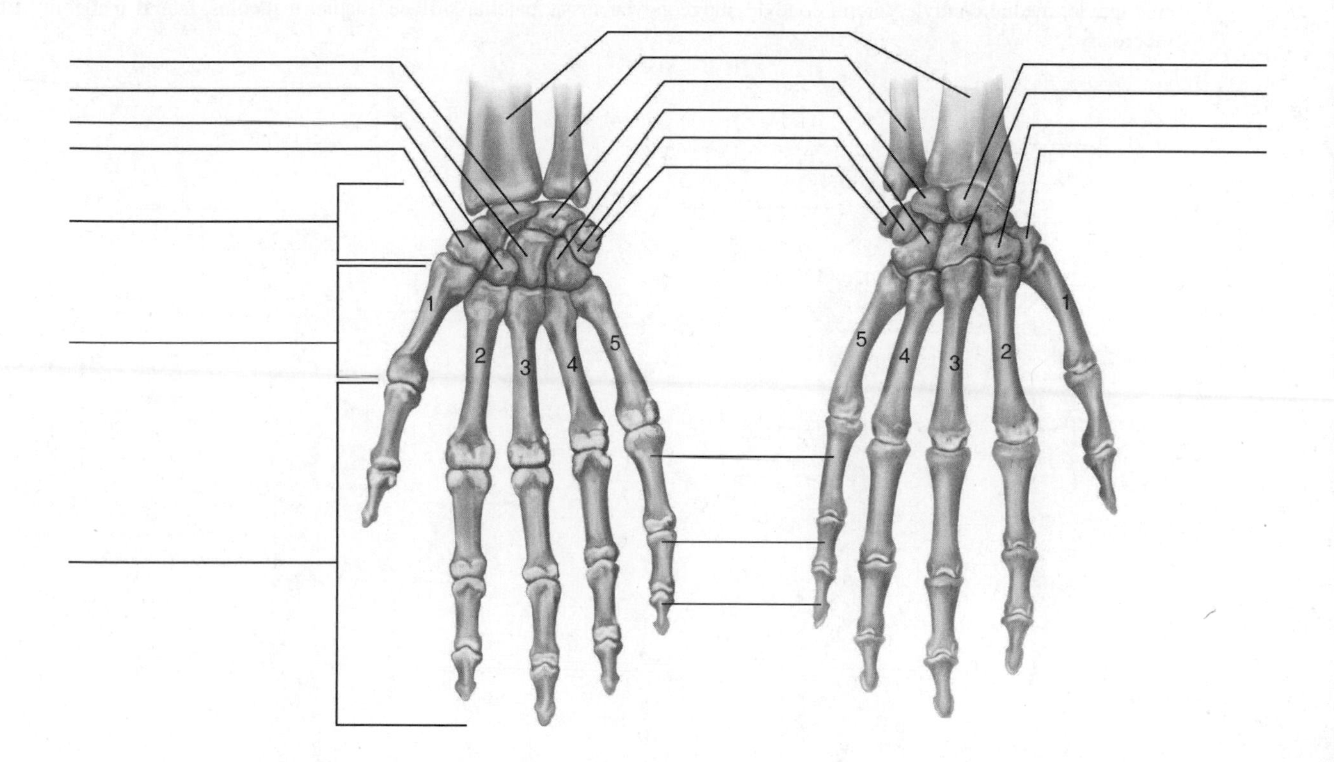

7.11 Pelvic Girdle (pp. 158–161)

A. List the bones of the pelvic girdle. (p. 158)

B. Identify the following structures in the bones of the pelvic girdle and assign a function or purpose to the structure (pp. 158–160): acetabulum, anterior superior iliac spine, ischial spine, obturator foramen

7.12 Lower Limb (pp. 161–163)

A. List the bones of the lower limb. (p. 161)

B. Identify the bone in which each of the following structures is located, and explain the function of each structure (pp. 161–162): head, neck, fovea capitis, gluteal tuberosity, greater trochanter, lesser trochanter, linea aspera, lateral epicondyle, medial epicondyle, medial condyle, lateral condyle, intercondylar fossa, patellar surface, medial malleolus, lateral malleolus, tibial tuberosity.

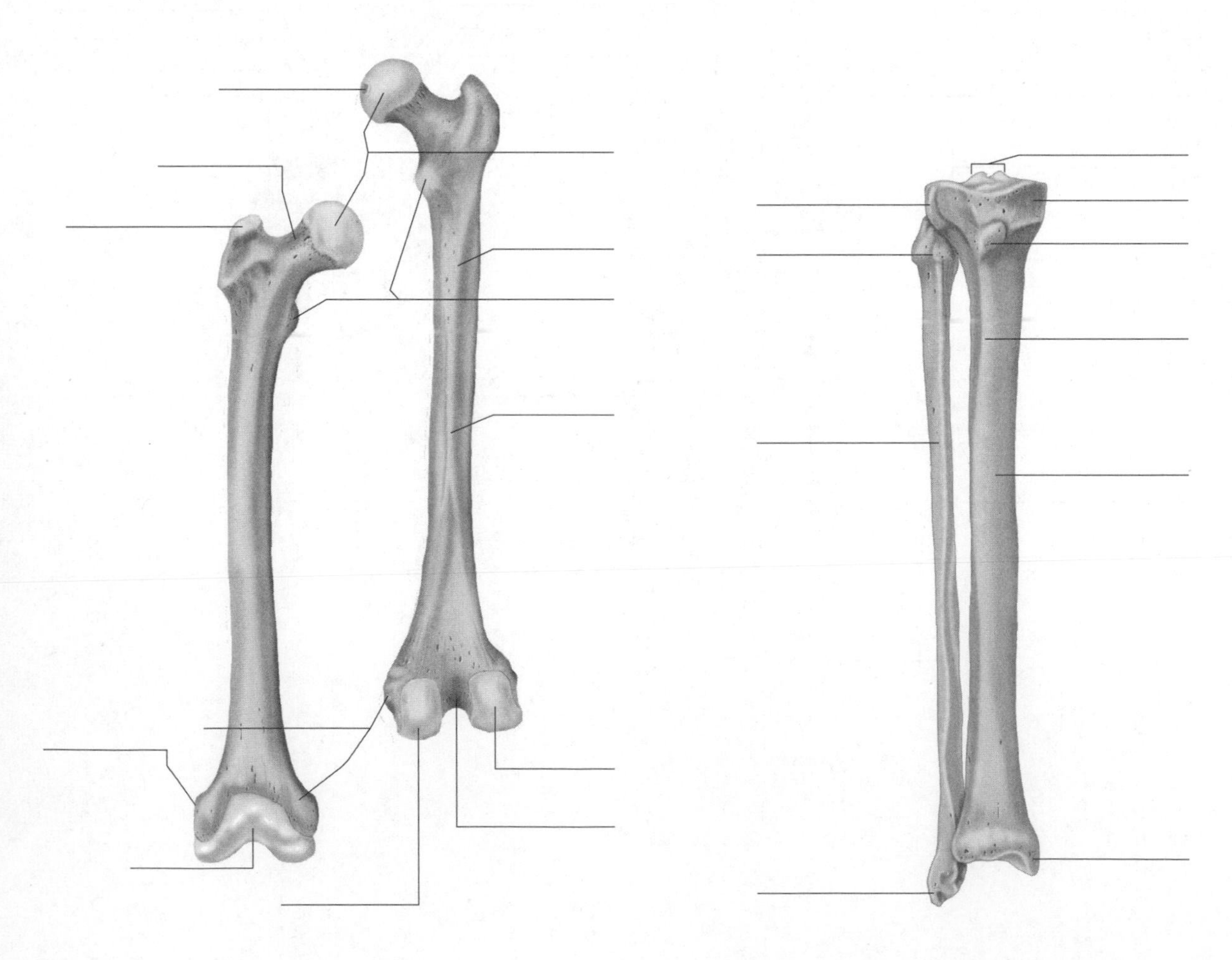

C. Label these structures on the accompanying illustration (p. 163): tarsal bones, calcaneus, talus, metatarsals, phalanges, navicular, cuboid, lateral cuneiform, intermediate cuneiform, medial cuneiform, proximal phalanx, middle phalanx, distal phalanx.

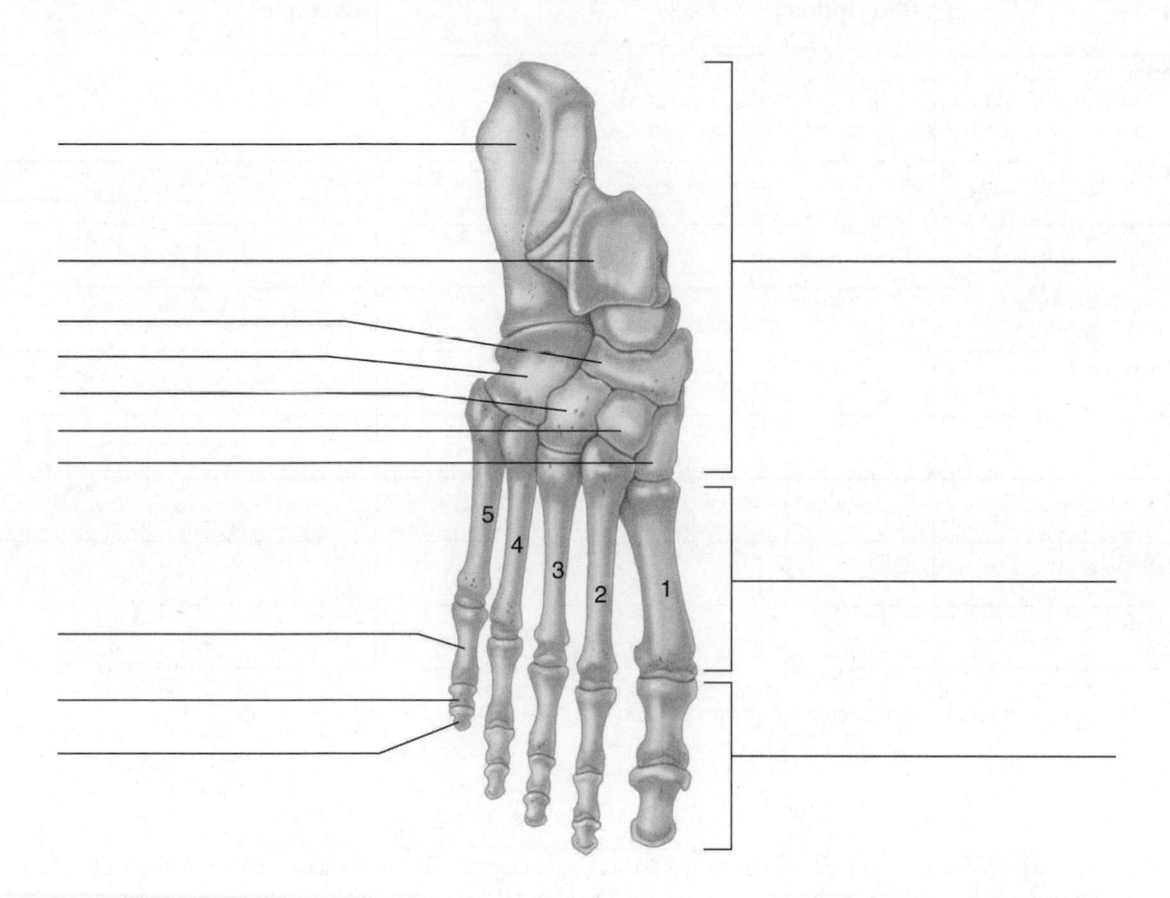

7.13 Joints (pp. 164–170)

A. Describe and give an example of each of the following types of joints. (pp. 165–167)

Joint	Description	Example
Fibrous joint		
Cartilaginous joint		
Synovial joint		
Ball-and-socket joint		
Condyloid joint		
Gliding joint		
Hinge joint		
Pivot joint		
Saddle joint		

B. Describe the structure and function of a synovial joint. (pp. 164–165)

C. Identify the following types of joint movements (pp. 167–170): turning the palms of the hands upward, shrugging the shoulders, bending the arm at the elbow, reaching for an object that is just beyond one's reach, turning the hands down so the palms face the floor, moving the legs apart in an "at ease" position, moving the legs together in an "attention" position, swiveling the head, drawing a large circle on a blackboard.

Clinical Focus Questions

A. Describe a lifestyle that may prevent the development of osteoporosis. Include the following.

1. Genetic factors
2. Nutritional factors
3. Types and amount of activity
4. Growth and development influences
5. Health education

B. Reflect on your own physical education experience and evaluate its impact on you.

When you have finished the study activities to your satisfaction, retake the mastery test and compare your results with your initial attempt. If you are not satisfied with your performance, repeat the appropriate study activities.

CHAPTER 8
MUSCULAR SYSTEM

OVERVIEW

In conjunction with the skeletal system, the muscular system moves the body. This chapter introduces the three types of muscle; the major events in contraction of skeletal, smooth, and cardiac muscles; the energy supply to muscle fiber for contraction; the occurrence of oxygen debt; and the process of muscle fatigue (learning outcomes 4, 5, 6, 7, and 8). This chapter also describes the structure and function of a skeletal muscle; distinguishes between a twitch and a sustained contraction (learning outcomes 2, 3, and 9); explains how various kinds of muscle contraction produce body movements and maintain posture (learning outcome 10); shows how the location and interaction of muscles produce body movements; identifies the location and action of major skeletal muscles (learning outcomes 14 and 15); and differentiates the structure and function of a multiunit smooth muscle and a visceral smooth muscle (learning outcome 11).

The skeletal system can be thought of as the passive partner in producing movement: think of the muscular system as the active partner. This chapter explains how muscles interact with bones to maintain posture and produce movement. In addition, it tells the characteristics and functions of skeletal, smooth, and cardiac muscles. This knowledge is a foundation for the study of other organ systems, such as the digestive, respiratory, and cardiovascular systems.

LEARNING OUTCOMES

After you have studied this chapter, you should be able to

8.1 Introduction

1. List various outcomes of muscular actions.

8.2 Structure of a Skeletal Muscle

2. Identify the structures that make up a skeletal muscle.
3. Identify the major parts of a skeletal muscle fiber and the function of each.
4. Discuss nervous stimulation of a skeletal muscle.

8.3 Skeletal Muscle Contraction

5. Identify the major events of skeletal muscle fiber contraction.
6. List the energy sources for muscle fiber contraction.
7. Describe how oxygen debt develops.
8. Describe how a muscle may become fatigued.

8.4 Muscular Responses

9. Distinguish between a twitch, recruitment, and a sustained contraction.
10. Explain how muscular contractions move body parts and help maintain posture.

8.5 Smooth Muscles

11. Distinguish between the structures and functions of multiunit smooth muscle and visceral smooth muscle.
12. Compare the contraction mechanisms of skeletal and smooth muscle fibers.

8.6 Cardiac Muscle

13. Compare the contraction mechanisms of cardiac and skeletal muscle fibers.

8.7 Skeletal Muscle Actions

14. Explain how the attachments, locations, and interactions of skeletal muscles make different movements possible.

8.8 Major Skeletal Muscles

15. Identify and locate the major skeletal muscles of each body region.
16. Identify the actions of the major skeletal muscles of each body region.

FOCUS QUESTION

How do muscle cells use energy and interact with bones to get you from your anatomy and physiology classroom to track team practice?

MASTERY TEST

Now take the mastery test. Do not guess. As soon as you complete the test, correct it. Note your successes and failures so that you can read the chapter to meet your learning needs.

1. The kind of energy that muscles use to contract is
   a. chemical.
   b. electrical.
   c. heat.
2. List the tissues found in skeletal muscle.
3. An individual skeletal muscle is separated from adjacent muscles by ____________________.
4. Layers of connective tissue extending into the muscle to form partitions between muscle bundles that are continuous with attachments of muscle to periosteum are called
   a. ligaments.
   b. fascia.
   c. aponeuroses.
   d. elastin.
5. The characteristic striated appearance of skeletal muscle is due to the arrangement of alternating protein filaments composed of____________________ and ____________________.
6. The functional units of muscle contraction are
   a. fascia.
   b. myofibrils.
   c. sarcomeres.
   d. troponin.
7. An injury in which a few muscle fibers are torn but the fascia is left intact is called a ____________________.
8. I bands are composed of ____________________.
9. The union between a nerve fiber and a muscle fiber is the
   a. motor neuron.
   b. motor end plate.
   c. neuromuscular junction.
   d. neurotransmitter.
10. A motor neuron and the muscle fibers it controls are called a ____________________.
11. When the cross-bridge of the myosin molecule forms linkages with actin filaments, the result is
    a. shortening of the muscle fiber.
    b. membrane polarization.
    c. release of acetylcholine.
12. The neurotransmitter that is necessary for the transmission of an impulse from a nerve to a skeletal muscle fiber is ______________.
13. The energy used in muscle contraction is supplied by the decomposition of ____________________.
14. A molecule that stores energy and can be used when ATP is in low supply is ____________________.
15. The ion necessary to allow linkage of myosin and actin is
    a. sodium.
    b. calcium.
    c. magnesium.
    d. potassium.
16. A person feels out of breath after vigorous exercise because of oxygen debt. Which of the following statements help(s) explain this phenomenon?
    a. Anaerobic respiration increases during strenuous activity.
    b. Lactic acid is metabolized more efficiently when the body is at rest.
    c. Conversion of lactic acid to glycogen occurs in the liver and requires energy.
    d. Priority in energy use is given to ATP synthesis.
17. After prolonged muscle use, muscle fatigue occurs due to an accumulation of ____________________.
18. The type of tissue that produces the major source of heat in the body is ____________________ tissue.
19. The minimal strength stimulus needed to elicit contraction of a single muscle fiber is called a(n) ____________________.

20. The strength of a muscle contraction in response to different levels of stimulation is determined by the
a. level of stimulation delivered to individual muscle fibers.
b. number of fibers that respond in each motor unit.
c. number of motor units stimulated.
d. characteristics of each muscle group.

21. The period of time between a stimulus to a muscle and muscle response is called the
a. latent period.
b. contraction.
c. refractory period.
d. relaxation

22. *Muscle tone* refers to
a. a state of sustained, partial contraction of muscles that is necessary to maintain posture.
b. a feeling of well-being following exercise.
c. the ability of a muscle to maintain contraction against an outside force.
d. the condition athletes attain after intensive training.

23. *Atrophy* refers to a(n) (increase, decrease) in the size and strength of a muscle.

24. Two types of smooth muscle are ______________________ muscle and ______________________ muscle.

25. Peristalsis is due to which of the following characteristics of smooth muscle?
a. capacity of smooth muscle fibers to excite each other
b. automaticity
c. rhythmicity
d. sympathetic innervation

26. Smooth muscle contracts (more slowly, more rapidly) than skeletal muscle following stimulation.

27. Impulses travel relatively (rapidly, slowly) through cardiac muscle.

28. The attachment of a muscle to a relatively fixed part is called the ______________________; the attachment to a relatively movable part is called the ______________________.

29. Smooth body movements depend upon ______________________ giving way to prime movers.

30. The muscle that compresses the cheeks inward when it contracts is the
a. orbicularis oris.
b. epicranius.
c. platysma.
d. buccinator.

31. The muscle that moves the head to one side is the
a. sternocleidomastoid.
b. splenius capitis.
c. semispinalis capitis.
d. longissimus capitis.

32. The muscle that abducts the upper arm and can both flex and extend the humerus is the
a. biceps brachii.
b. deltoid.
c. infraspinatus.
d. triceps brachii.

33. The band of tough connective tissue that extends from the xiphoid process to the symphysis pubis and serves as an attachment for muscles of the abdominal wall is the ______________________.

34. The heaviest muscle in the body, which straightens the leg at the hip during walking, is the
a. psoas major.
b. gluteus maximus.
c. adductor longus.
d. gracilis.

STUDY ACTIVITIES

Aids to Understanding Words

Define the following word parts. (p. 179)

calat-

erg-

hyper-

inter-

laten-

myo-

sarco-

syn-

tetan-

-troph

8.1 Introduction (p. 179)

List the three types of muscle.

8.2 Structure of a Skeletal Muscle (pp. 179–182)

A. List the kinds of tissue present in skeletal muscle. (p. 179)

B. Answer these questions concerning skeletal muscle tissues. (p. 179)

1. A skeletal muscle is held in position by layers of fibrous connective tissue called ________________________.
2. This tissue extends beyond the end of a skeletal muscle to form a cordlike ________________________.
3. When this tissue extends beyond the muscle to form a sheetlike structure, it is called a(n) ________________.
4. What are fascicles?
5. The connective tissue that covers each muscle fiber within a fasicle is the ________________.

C. Answer these questions concerning skeletal muscle fibers. (pp. 179–182)

1. Describe a single muscle fiber.
2. Describe the structure and function of a sarcomere.
3. The network of membranous channels in the cytoplasm of muscle fibers is the ____________________________. The other channels found in the cytoplasm are the ______________________.
4. What is the function of these channels?

D. Label these structures in the accompanying illustration of a neuromuscular junction (p. 182): mitochondria, synaptic cleft, synaptic vesicles, folded sarcolemma, muscle fiber nucleus, myofibril of muscle fiber, motor end plate, (2) axon branches, motor neuron axon.

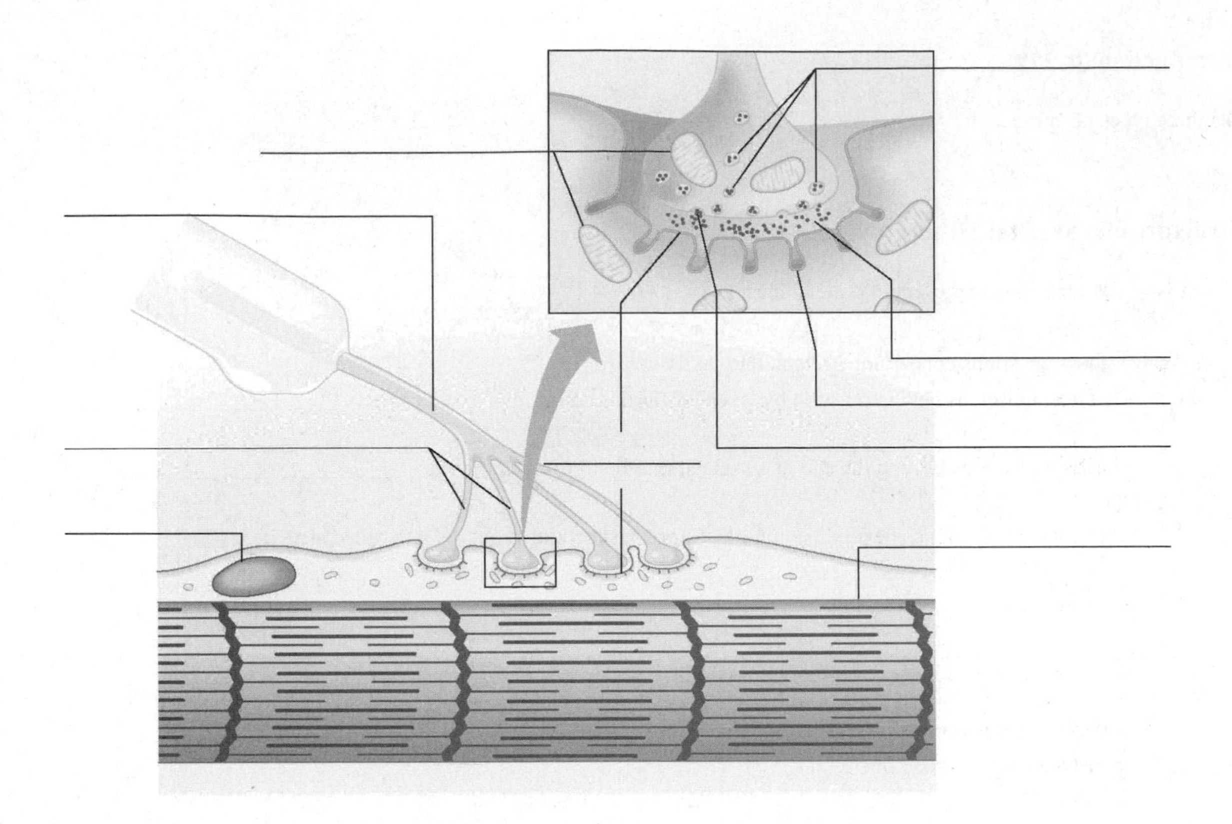

8.3 Skeletal Muscle Contraction (pp. 182–187)

A. Describe the roles of actin and myosin in muscle contraction. (p. 183)

B. Describe the transmission of a nerve impulse across the neuromuscular junction. (p. 183)

C. Answer these questions concerning stimulus for contraction. (pp. 183–184)

1. Describe the interaction of acetylcholine and calcium ions in stimulating muscle contraction.

2. What is a cross-bridge?

3. The action of acetylcholine is halted by the enzyme _________________________.

D. Answer these questions concerning energy sources for contraction. (pp. 184–186)

1. How does ATP supply energy for muscle contraction?

2. How does creatine phosphate supply energy for muscle contraction?

E. Answer these questions concerning oxygen supply and cellular respiration. (pp. 186 187)

1. What substance in muscle seems able to store oxygen temporarily?

2. Why is oxygen necessary for muscle contraction?

3. How does the muscle continue to contract in the absence of oxygen?

4. What is meant by oxygen debt?

F. Answer these questions concerning muscle fatigue. (p. 187)

1. What is meant by muscle fatigue? What causes it?

2. Less than half the energy released by cellular respiration is available for metabolic processes. The rest is lost as ________________________.

8.4 Muscular Responses (pp. 187–191)

A. Answer these questions concerning muscular responses. (pp. 187–188)

1. Define *threshold stimulus*.

2. What happens once the threshold stimulus is reached?

B. The accompanying illustration shows a myogram of a type of muscle contraction known as a twitch. Label the following: time of stimulation, latent period, period of contraction, period of relaxation. Explain the significance of each of these events. (p. 188–189)

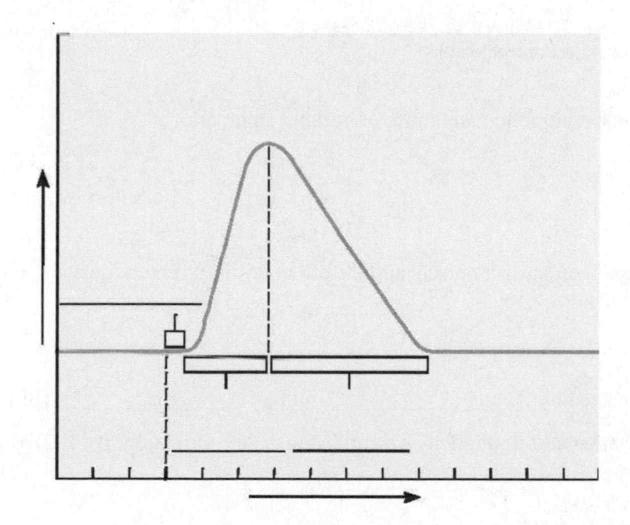

C. Describe the role of slow muscle fibers and fast muscle fibers in the response of a muscle to different types of exercise. (p. 189)

D. Describe these kinds of muscle contractions. (p. 190)

1. Summation

2. Recruitment of motor units

3. Sustained contraction

4. Muscle tone

8.5 Smooth Muscles (p. 191)

A. Compare the structure of smooth muscle fibers and skeletal muscle fibers. (p. 191)

B. Answer these questions about smooth muscle. (p. 191)

1. The two types of smooth muscle are ____________________ smooth muscle and ____________________ smooth muscle.

2. Where are each of these types of smooth muscle found?

3. The properties of smooth muscle that allow peristalsis are ______________ and ______________.

C. Answer these questions about smooth muscle contraction. (p. 191)

1. The neurotransmitters that affect smooth muscle are____________________ and ______________.

2. Do these neurotransmitters always lead to muscle contraction?

3. Compare the characteristics of smooth muscle contraction and those of skeletal contraction.

8.6 Cardiac Muscle (pp. 191–192)

A. Describe the structure of the sarcoplasmic reticulum of cardiac muscle and the effect of this structure on cardiac muscle contraction. (p. 192)

B. Opposing ends of cardiac muscle fibers are connected by______________________________. (p. 192)

C. Describe the effect of the properties of self-excitation and rhythmicity on cardiac muscle contraction. (p. 192)

Genetics Connection 8.1 (p. 193)

Match the following descriptions with the genetic diseases they describe. (p. 193)

_____ a. muscular dystrophies

_____ b. Charcot-Marie-Tooth disease

_____ c. myotonic dystrophy

_____ d. hereditary idiopathic dilated cardiomyopathy

1. delayed muscle reaction following contraction in which the symptoms become increasingly severe in successive generations
2. a genetic error that leads to the change of a single DNA nucleotide base leading to a defect in the ability of actin to anchor to Z bands.
3. muscles that weaken and degenerate due to abnormalities of the protein dystrophin
4. characterized by progressing weakness in the muscles of the hands and feet diagnosed by electromyography and nerve conduction velocity studies

8.7 Skeletal Muscle Actions (pp. 192–194)

A. A skeletal muscle has at least two places of attachment to bone. For instance, the gluteus maximus, which extends the leg at the hip, is attached to the posterior surface of the ilium, the sacrum, and the coccyx at one end and to the posterior surface of the femur and the iliotibial tract at the other. One place of muscle attachment is the origin, and the other is the insertion. Explain the difference between the two. (p. 192)

B. Match the terms in the first column with the statements in the second column that best describe the role of muscle groups in producing smooth muscle movement. (p. 194)

_____ a. prime mover

_____ b. synergist

_____ c. antagonist

1. muscle that returns a part to its original position
2. muscle that makes the action of the prime mover more effective
3. muscle that has the major responsibility for producing a movement

C. Define *flexion* and *extension.* (pp. 192–193)

8.8 Major Skeletal Muscles (pp. 194–207)

A. Identify and locate the major skeletal muscles of each body region. Check with the instructor regarding which muscles to focus on.

B. Identify the actions of the major skeletal muscles of each body region. Check with the instructor regarding which muscles to focus on.

Clinical Focus Questions

You are a school nurse in a high school known for the excellence of its athletic programs. You have been asked by the board of education to address the issue of steroid use by male and female athletes. In preparing your statement, you must address the following issues.

1. Why do athletes use steroid drugs?
2. What kind of testing program can be implemented to ensure that students in this school are not using such drugs?
3. What are the symptoms of steroid use in males and females?
4. What problems are associated with use of these drugs?

When you have finished the study activities to your satisfaction, retake the mastery test and compare your results with your initial attempt. If you are not satisfied with your performance, repeat the appropriate study activities.

UNIT 3

INTEGRATION AND COORDINATION

CHAPTER 9

NERVOUS SYSTEM

OVERVIEW

The human body depends upon the nervous system and the endocrine system to coordinate and integrate the functions of other body systems so that the internal environment remains able to function normally and the body can respond to its environment. This chapter focuses on the nervous system. The structure and function of the various parts of the nervous system and its various tissues are the foundation for your understanding of the nervous system (learning outcomes 1–6). You will learn how neurons of various types are classified and how they function (learning outcomes 7 and 8). You will study how changes in membrane potential are related to excitatory and inhibitory neurotransmitters as the nervous system processes information (learning outcomes 9–14). The classification of nerve fibers in peripheral nerves is discussed, along with the structure and function of nerve pathways (learning outcomes 15 and 16). Finally, the function and structure of the brain, spinal cord, and their coverings are addressed, as well as the functions of the peripheral and autonomic nervous systems and their various divisions (learning outcomes 17–27).

LEARNING OUTCOMES

After you have studied this chapter, you should be able to

9.1 Introduction

1. Distinguish between the two types of cells that compose nervous tissue.
2. Name the two major groups of nervous system organs.

9.2 General Functions of the Nervous System

3. Explain the general functions of the nervous system.

9.3 Neuroglia

4. State the functions of neuroglia in the central nervous system.
5. Distinguish among the types of neuroglia in the central nervous system.
6. Describe the Schwann cells of the peripheral nervous system.

9.4 Neurons

7. Describe the general structure of a neuron.
8. Explain how differences in structure and function are used to classify neurons.

9.5 The Synapse

9. Explain how information passes from one neuron to another.

9.6 Cell Membrane Potential

10. Explain how a membrane becomes polarized.
11. Describe the events that lead to the generation of an action potential.

9.7 Nerve Impulses

12. Compare nerve impulse conduction in myelinated and unmyelinated neurons.

9.8 Synaptic Transmission

13. Identify the changes in membrane potential associated with excitatory and inhibitory neurotransmitters.

9.9 Impulse Processing

14. Describe the general ways in which the nervous system processes information.

9.10 Types of Nerves

15. Describe how peripheral nerves are classified.

9.11 Nerve Pathways

16. Describe the function of each part of a reflex arc, and name two reflex examples.

9.12 Meninges

17. Describe the coverings of the brain and spinal cord.

9.13 Spinal Cord

18. Describe the structure of the spinal cord and its major functions.

9.14 Brain

19. Name the major parts and functions of the brain.
20. Distinguish among motor, sensory, and association areas of the cerebral cortex.
21. Describe the location, formation, and function of cerebrospinal fluid.

9.15 Peripheral Nervous System

22. List the major parts of the peripheral nervous system.
23. Name the cranial nerves and list their major functions.
24. Describe the structure of a spinal nerve.

9.16 Autonomic Nervous System

25. Describe the functions of the autonomic nervous system.
26. Distinguish between the sympathetic and parasympathetic divisions of the autonomic nervous system.
27. Describe a sympathetic and a parasympathetic nerve pathway.

FOCUS QUESTION

It is noon, and you are just finishing an anatomy assignment. You hear your stomach growling and realize you are hungry. You make a sandwich and pour a glass of milk. After you finish eating, you realize you have been studying for three hours and should go for a walk. How does the nervous system receive internal and external cues, process incoming information, and decide what action to take?

MASTERY TEST

Now take the mastery test. Do not guess. As soon as you complete the test, correct it. Note your successes and failures so that you can read the chapter to meet your learning needs.

1. The basic unit of structure and function of the nervous system is the ______________________.

2–5. Match the function listed in the left column with the structure in the right column.

Function	Structure
a. information transmitted in the form of electrical and chemical changes	2. ____ axon
b. extension of neuron cell body	3. ____ nerve impulse
c. receives information as electrochemical messages	4. ____ nerve fiber
d. sends information	5. ____ dendrite

6. The three general functions of the peripheral nervous system are ______________________, ______________________, and ______________________.

7. The motor functions of the nervous system that are consciously controlled are part of the ______________________ nervous system.

8. The supporting cells of the central nervous system are ______________________ cells.

9. The cells of the central nervous system that divide throughout life are the ______________________ cells.

10. The structures that are primarily responsible for the "blood–brain" barrier are the ______________________ of the brain and the glial cells known as ______________________.

11. List the common features of all neurons.

12. Axons arise from a part of the cell body known as the ______________________.

13. The neurilemma is composed of
a. Nissl bodies.
b. myelin.
c. the cytoplasm and nuclei of Schwann cells.
d. neuron cell bodies.

14. List the major groups of neurons classified on the basis of structure.

15. Neurons may be classified functionally as ____________________, ____________________, and ________________.

16. When the nerve cell is at rest, the concentration of ____________________ ions is relatively greater on the outside of the cell membrane.

17. When the threshold potential is reached, the region of the cell membrane being stimulated undergoes a change in ____________________.

18. The rapid sequence of depolarization and repolarization in the nerve cell is known as the ____________________ ____________________.

19. Nerves with ______________ diameters conduct impulses faster than those with ______________ diameters.

20. The junction between two communicating neurons is called a(n) ______________________.

21. Transmission of nerve impulses from one neuron to another is controlled by substances called ______________________.

22. Neurotransmitters can either inhibit or excite nerve transmission.
a. True
b. False

23. The function of neuronal pools is ______________________ of nerve impulses.

24. Divergence occurs when
a. an impulse is amplified.
b. an additive effect is noted.
c. an impulse reaches different regions of the brain.
d. nerve fibers divide.

25. A bundle of nerve fibers held together by connective tissue is a(n) ______________________.

26. An automatic, unconscious response to a change inside or outside the body is a(n) ______________________.

27. The outer membrane covering the brain is composed of fibrous connective tissues and is called the
a. dura mater.
b. arachnoid mater.
c. pia mater.
d. periosteum.

28. Cerebrospinal fluid is found between the
a. arachnoid mater and the dura mater.
b. vertebrae and the meninges.
c. pia mater and the arachnoid mater.

29. The spinal cord ends
a. at the sacrum.
b. between thoracic vertebrae 11 and 12.
c. between lumbar vertebrae 1 and 2.
d. at lumbar vertebra 5.

30. Which of the following statements is true about the white matter in the spinal cord?
a. A cross section of the cord reveals a core of white matter surrounded by gray matter.
b. The white matter is composed of myelinated nerve fibers and makes up nerve pathways called tracts.
c. The white matter carries sensory stimuli to the brain; the gray matter carries motor stimuli to the periphery.
d. The nerve fibers within spinal tracts arise from cell bodies located in the same part of the nervous system.

31. The four major portions of the brain are the ______________ ______________, ______________, and ______________.

32. The hemispheres of the cerebrum are connected by nerve fibers called the
a. corpus callosum.
b. falx cerebri.
c. tissue of Rolando.
d. tentorium.

33. Match the functions in the first column with the appropriate area of the brain in the second column.

_____a. hearing
_____b. vision
_____c. recognition of printed work
_____d. control of voluntary muscles
_____e. pain
_____f. complex problem solving

1. frontal lobes
2. parietal lobes
3. temporal lobes
4. occipital lobes

34. Which hemisphere of the brain is dominant for most of the population?

35. Cerebrospinal fluid is produced by the ______________________________________.

36. The thalamus and hypothalamus are parts of the brain located in the
a. midbrain.
b. pons.
c. medulla oblongata.
d. diencephalon.

37. The part of the brain responsible for regulation of temperature and heart rate, control of hunger, and regulation of fluid and electrolytes is the
a. thalamus.
b. hypothalamus.
c. medulla oblongata.
d. pons.

38. The ______________________________________ produces emotional reactions of fear, anger, and pleasure.

39. Consciousness is dependent upon stimulation of the ____.

40. Tremors, loss of muscle tone, gait disturbance, and a loss of equilibrium may be due to damage to the ________________________.

41. The peripheral nervous system has two divisions: the ___________________________ nervous system and the ______________________________ nervous system.

42. There are __________________________ pairs of cranial nerves; all but one of these arise from the ___________________________.

43. Vision and function of the eyes and associated structures are controlled by cranial nerves______ through______.

44. There are ____________pairs of spinal nerves.

45. The part of the nervous system that functions without conscious control is the _______________________ nervous system.

46. Nerves of the sympathetic division leave the spinal cord through the _________________________ of spinal nerves in the ___________________________ through the ______________________________ segments.

47. Which of the following are responses to stimulation by the sympathetic nervous system?
a. increased heart rate
b. increased blood glucose concentration
c. increased peristalsis
d. increased salivation

48. Which of the following are responses to stimulation of the parasympathetic nervous system?
a. dilation of the bronchioles
b. dilation of the coronary arteries
c. contraction of the gallbladder
d. contraction of the muscles of the urinary bladder

STUDY ACTIVITIES

Aids to Understanding Words

Define the following word parts. (p. 214)

ax-

dendr-

funi-

gangli-

-lemm

mening-

moto-

peri-

plex-

sens-

syn-

ventr-

9.1 Introduction (p. 214)

A. The structural and functional units of the nervous system are the ____________________.

B. Neurons may have many ______________________ and ________________ but typically have only one ________________.

C. List the organs of the central and peripheral nervous systems.

D. What does the nervous system allow us to do?

9.2 General Functions of the Nervous System (pp. 215–216)

A. Describe the general functions of the nervous system: sensory, integrative, and motor.

B. Describe the somatic and autonomic nervous systems.

9.3 Neuroglia (p. 216)

A. Label the following structures in the accompanying illustration (p. 217): ependymal cell, astrocyte, neurons, oligodendrocyte, microglial cells, capillary, axon, myelin sheath (cut), node.

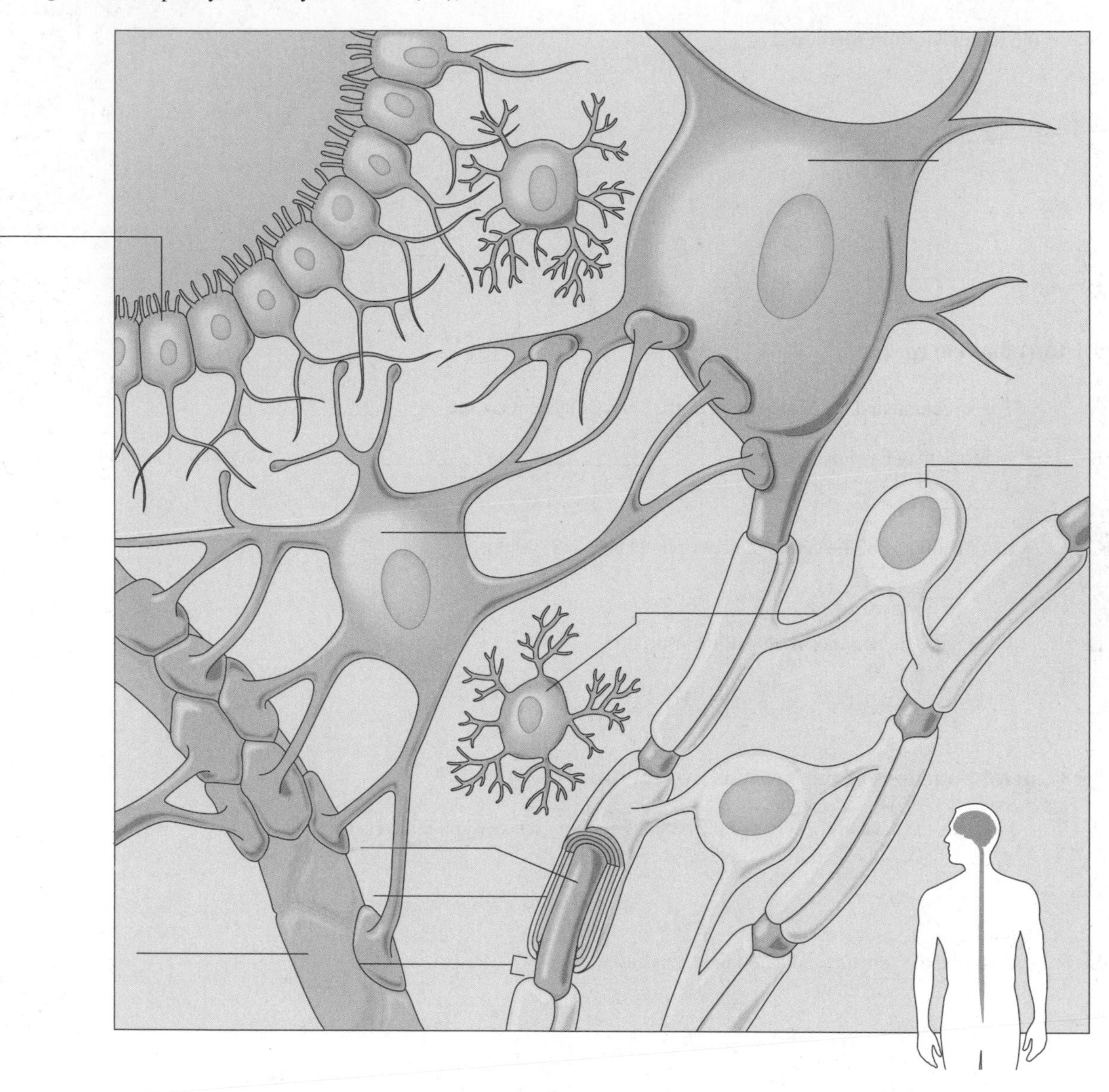

B. Fill in the following table comparing supporting cells of the nervous system.

Neuroglia	Location	Function
Microglial cells		
Oligodendrocytes		
Astrocytes		
Ependymal cells		
Schwann cells		

9.4 Neurons (pp. 216–221)

A. Label the following structures of a motor neuron shown in the accompanying illustration (p. 218): dendrites, axon, nucleolus, cell body, nucleus, neurofibrils, nodes of Ranvier, myelin, nucleus of Schwann cell, axonal hillock, chromatophilic substance, schwann cell synaptic knob of axon terminal, portion of a collateral axon.

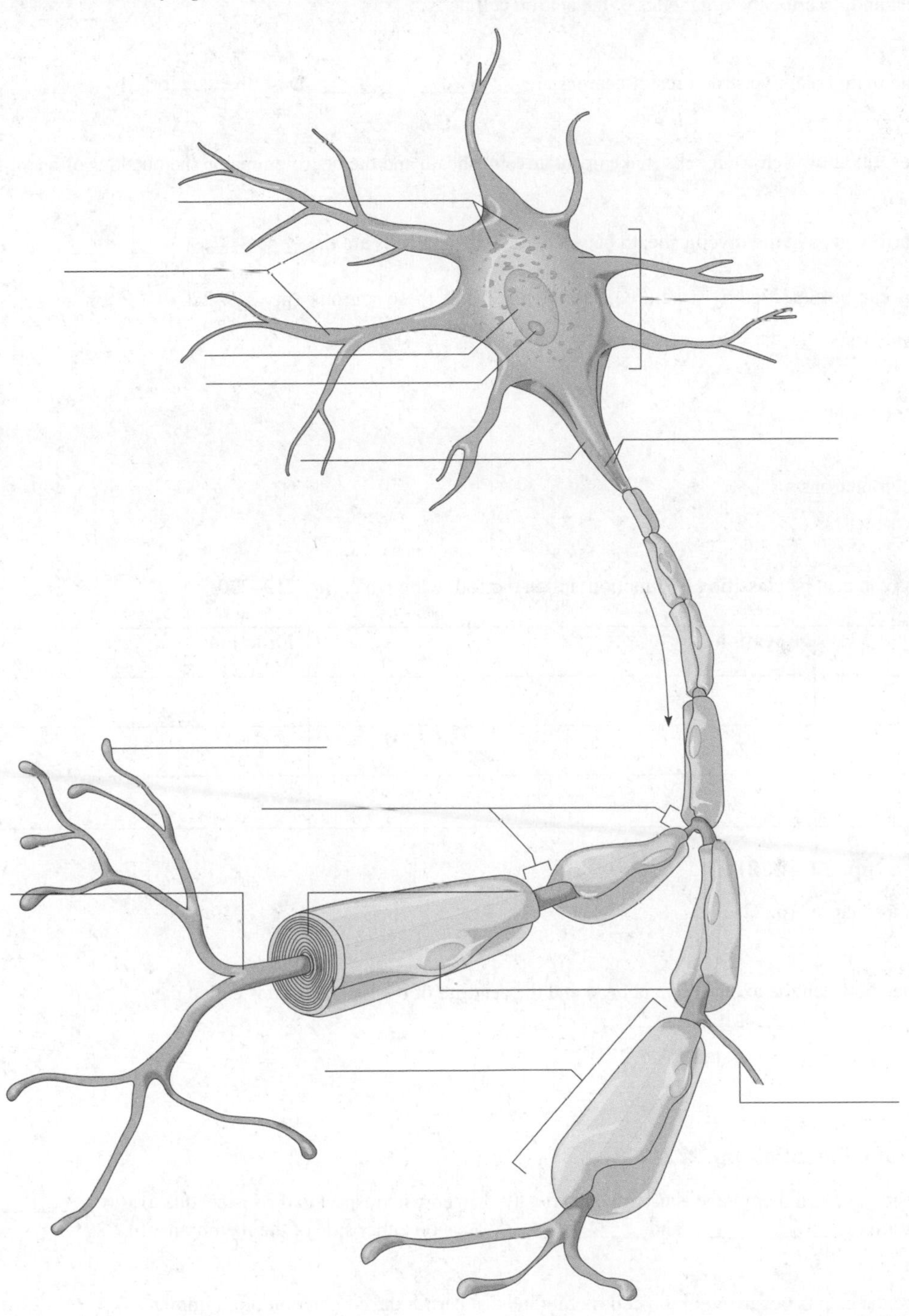

B. Answer these questions about neuron structure. (pp. 216–218)

1. List the basic structures common to all neurons.

2. List and describe the organelles of the neuron cell body.

3. The main receptive structures of neurons are _______________. Describe them briefly.

4. Describe how Schwann cells make up the myelin sheath and the neurolemma on the outsides of nerve fibers.

5. Narrow gaps in the myelin sheath between the Schwann cells are the ____________________________.

C. Neurons can be classified by structure. Describe and locate these neurons. (pp. 219–220)

bipolar neurons

unipolar neurons

multipolar neurons

D. Neurons can also be classified by function. Fill in the following table. (pp. 219–220)

Neuron	Location	Function
Sensory neurons		
Interneurons		
Motor neurons		

9.5 The Synapse (pp. 221–222)

A. What is a synapse? (p. 221)

B. The space between the axon of one neuron and the dendrite of the next neuron is called the ____________________________.

C. Describe synaptic transmissions. (pp. 221–222)

9.6 Cell Membrane Potential (pp. 222–227)

A. The outside of a cell membrane is usually electrically charged with respect to the inside, due to a(n) ________________ distribution of ____________ and ____________ ions on either side of the membrane. (p. 222)

B. Describe the events occurring in the cell membrane that permit the conduction of an impulse. (pp. 222–225)

C. Describe membrane polarization, depolarization, and repolarization. Which of these events is a nerve impulse? (pp. 222–226)

9.7 Nerve Impulses (pp. 227–228)

A. How do the nodes of Ranvier affect nerve impulse conduction? What kind of conduction is this called? (p. 227)

B. Define the *all-or-none response* in neurons. (pp. 227–228)

C. What is the refractory period? (p. 228)

9.8 Synaptic Transmission (p. 228)

A. Answer the following questions regarding synaptic transmission.

1. Label the following structures in the accompanying drawing of a synapse (p. 222): synaptic cleft, neurotransmitter, axon membrane, synaptic knob, vesicle-releasing neurotransmitter.

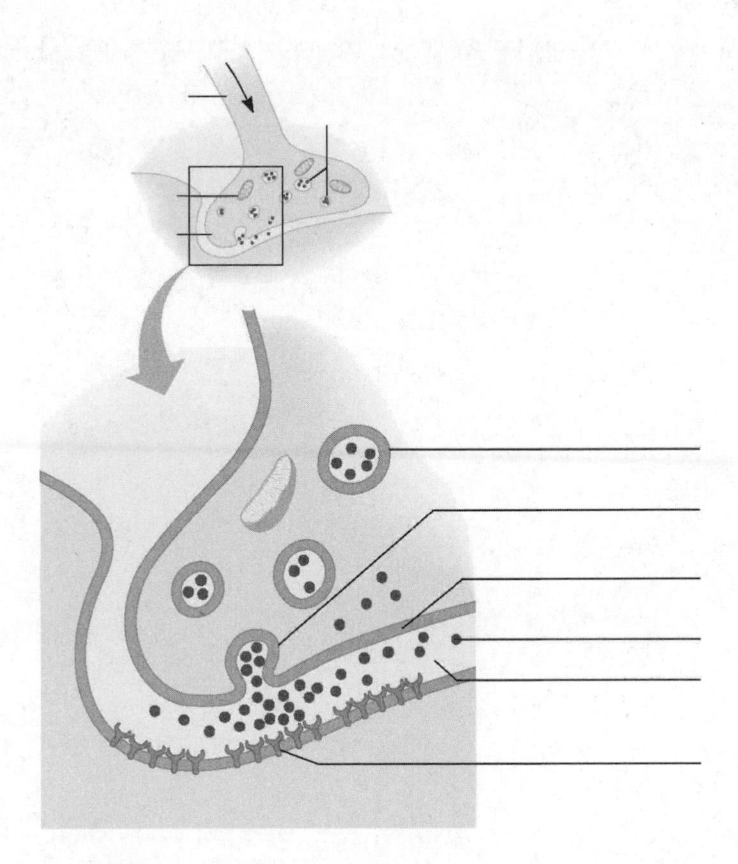

2. How does a neurotransmitter initiate depolarization?

3. List substances that function as neurotransmitters.

B. Answer these questions concerning excitatory and inhibitory actions. (p. 228)

1. Describe excitatory and inhibitory actions. How do they interact in normal nerve function?

2. What substances seem to have inhibitory action? What substances have excitatory action?

3. What ion stimulates the release of neurotransmitters at the synapse?

4. How is stimulation of the nerve fiber stopped?

9.9 Impulse Processing (pp. 228–230)

A. Describe the role of neuronal pools in producing facilitation, convergence, and divergence. (pp. 228–230)

B. What is facilitation? (p. 229)

C. What is convergence? (p. 229)

D. What is divergence? (pp. 229–230)

9.10 Types of Nerves (p. 230)

A. What is a nerve?

B. What is a motor nerve?

C. What is a sensory nerve?

D. What is a mixed nerve?

9.11 Nerve Pathways (pp. 231–232)

A. What is a nerve pathway? (p. 231)

B. What is a reflex? What is a reflex arc? (p. 231)

C. Label the following parts of the reflex shown in the drawing (p. 231): cell body of sensory neuron, spinal cord, direction of impulse, receptor associated with dendrites of sensory neuron, axon of sensory neuron, cell body of motor neuron, axon of motor neuron, effector (quadriceps femoris muscle) group, patella, patellar ligament.

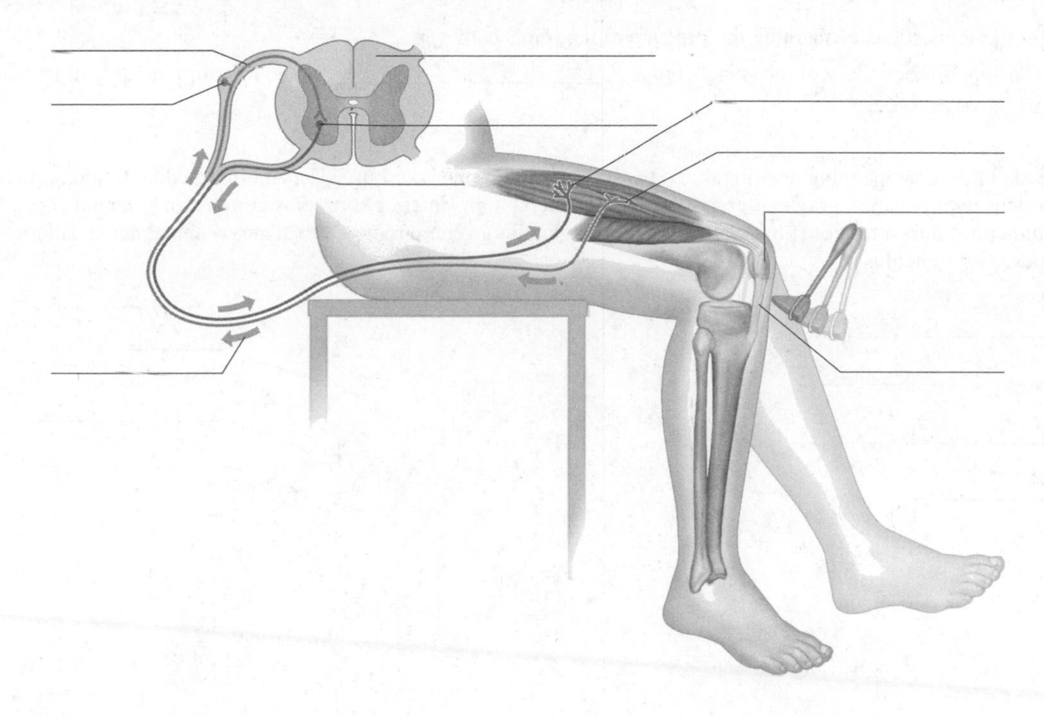

9.12 Meninges (pp. 232–234)

A. What are the layered membranes of the central nervous system that lie between bony covering and soft tissues called? What are the functions of these membranes? (pp. 232–233)

B. Fill in the following table regarding the meninges of the central nervous system. (pp. 233–234)

Layer	Location	Structure and special features	Function
Dura mater			
Arachnoid mater			
Pia mater			

9.13 Spinal Cord (pp. 234–236)

A. Answer these questions concerning the structure of the spinal cord. (pp. 234–235)

1. The superior boundary of the spinal cord is __________________________. The inferior boundary of the spinal cord is _____________________________.

2. Label the accompanying drawing of a cross section of the spinal cord (p. 235): anterior median fissure, posterior median sulcus, white matter, gray matter, posterior horn, lateral horn, anterior horn, gray commissure, central canal, lateral funiculus, dorsal root of spinal nerve, dorsal root ganglion, ventral root of spinal nerve, spinal nerve, anterior funiculus, posterior funiculus.

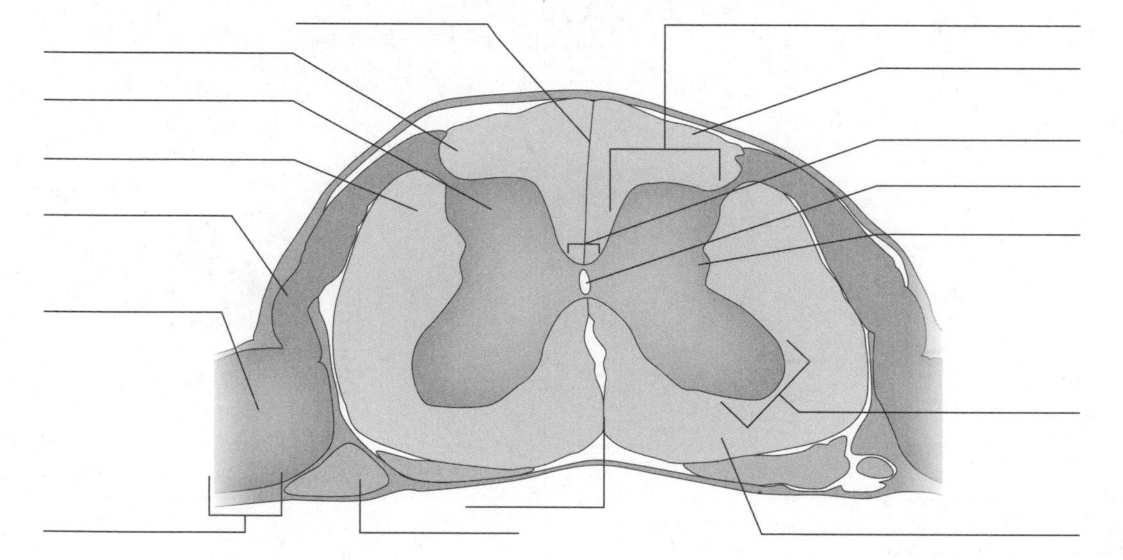

3. There are __________ pairs of spinal nerves.

B. Describe the two-way communication system between the brain and the body parts that delivers information to the brain and carries information from the brain. Include the names of the ascending and descending tracts as determined by their initiation and termination. (pp. 234–235)

9.14 Brain (pp. 236–246)

A. List the parts of the brain. (p. 236)

B. Answer these questions concerning the structure of the cerebrum. (pp. 236–238)

1. The bridge that connects the two hemispheres is the ______________________.
2. The ridges of the hemispheres are called ______________________.
3. A shallow groove is called a(n) ______________________; a deeper groove is called a(n) ______________________.
4. Name the lobes of the cerebral hemispheres.
5. The outer layer of the cerebrum is the ______________________. It is composed of ______________________.
6. Just beneath this outer layer of the cerebrum is the ______________________. It is composed of ______________________.

C. Answer these questions concerning the function of the cerebrum. (pp. 238–239)

1. What are the functions of the cerebrum?
2. Fill in the following table regarding the functional areas of the cerebrum.

Area	**Locations**	**Function**
Motor		
Sensory		
Association		

D. Describe the functions and locations of the dominant and nondominant hemispheres of the brain. (pp. 239–240)

E. List the basal ganglia and describe their functions. (p. 240)

F. Answer these questions concerning the ventricles and cerebrospinal fluid. (pp. 240–242)

1. Describe the location of the four ventricles.

2. Where is cerebrospinal fluid secreted? What is its function?

G. Answer these questions concerning the diencephalon. (pp. 242–243)

1. Locate the diencephalon and describe its structure.

2. Fill in the following table regarding the functions of the diencephalon.

Structure	Location	Function
Thalamus		
Hypothalamus		

3. What structures constitute the limbic system? What is the function of this system?

H. Answer these questions concerning the brainstem. (pp. 243–244)

1. Where is the brainstem? What are its component structures?

2. Where is the midbrain located? What is its function?

3. Where is the pons located? What is its function?

4. Where is the medulla oblongata located? The medulla is the control center for what vital activities?

5. Describe the location, structure, and function of the reticular formation.

I. Describe the cerebellum and outline its function as an integrator and coordinator of sensory and motor function. (pp. 244–246)

9.15 Peripheral Nervous System (pp. 246–250)

A. Answer these questions concerning the parts of the peripheral nervous system. (p. 246)

1. What are the parts of the peripheral nervous system?

2. What is the function of the somatic nervous system? The autonomic nervous system?

B. Answer these questions concerning the cranial nerves. (pp. 246–248)
You need to identify the cranial nerves by *both the name and number*. Fill in the following table.

Cranial nerve	Sensory, motor, or mixed	Function
Olfactory I		
Optic II		
Oculomotor III		
Trochlear IV		
Trigeminal V		
Abducens VI		
Facial VII		
Vestibulocochlear VIII		
Glossopharyngeal IX		
Vagus X		
Accessory XI		
Hypoglossal XII		

There are many mnemonic devices for the cranial nerves, here are two suggestions:

On (I) old (II) Olympus (III) towering (IV) top (V) a (VI) finely (VII) vested (VIII) German (IX) viewed (X) a (XI) hawk (XII).

On (I) occasion (II) our (III) trusty (IV) truck (V) acts (VI) funny (VII) very (VIII) good (IX) vehicle (X) any (XI) how (XII).

C. Answer these questions concerning spinal nerves. (pp. 248–250)

1. How are the spinal nerves identified?

2. Describe the structure and function of the dorsal root and the ventral root.

D. Answer these questions concerning the spinal nerve plexuses. (pp. 249–250)

1. What is a plexus?

2. Fill in the following table regarding the spinal nerve plexuses.

Plexus	Nerves involved	Structures innervated
Cervical plexuses		
Brachial plexuses		
Lumbosacral plexuses		

9.16 Autonomic Nervous System (pp. 250–254)

A. Describe the characteristics of the autonomic nervous system and compare it to the somatic motor nervous system. (pp. 250–251)

B. What are the two divisions of the automonic nervous system? Describe structures of both divisions; include the preganglionic, ganglia, and postganglionic structures. (p. 251)

C. Identify the neurotransmitters associated with preganglionic and postganglionic fibers in the sympathetic and parasympathetic divisions and what functions are controlled or influenced by them. (pp. 251–254)

Clinical Focus Questions

Multiple sclerosis is a disease in which myelin is destroyed, leading to the formation of plaques or scars in the central nervous system. The course of the disease is characterized by periods of remission and exacerbation.

A. How might nerve transmission be affected?

B. Remission may be due to the replacement of myelin. Which of the glial cells would be involved in this process?

C. The symptoms of multiple sclerosis depend upon the location and duration of the characteristic lesions. Predict the expected symptoms if the lesion is located in the following areas:

optic nerves

brainstem

cerebellum

corticospinal tracts

white matter of the cerebral cortex

When you have finished the study activities to your satisfaction, retake the mastery test and compare your results with your initial attempt. If you are not satisfied with your performance, repeat the appropriate study activities.

CHAPTER 10

THE SENSES

OVERVIEW

This chapter deals with specialized parts of the nervous system that allow the body to assess and adjust to both the external and internal environments. It describes the locations and structures of the somatic (learning outcomes 1–5) and special senses (learning outcomes 6–14), as well as the function of each in maintaining homeostasis.

An understanding of these senses is necessary to know how the nervous system receives input and responds to support life.

LEARNING OUTCOMES

After you have studied this chapter, you should be able to

10.1 Introduction

1. Distinguish between general senses and special senses.

10.2 Receptors, Sensations, and Perception

2. Name five kinds of receptors, and explain their functions.
3. Explain how a sensation arises.

10.3 General Senses

4. Describe the receptors associated with the senses of touch, pressure, temperature, and pain.
5. Describe how the sense of pain is produced.

10.4 Special Senses

6. Identify the locations of the receptors associated with the special senses.

10.5 Sense of Smell

7. Explain the relationship between the senses of smell and taste.
8. Explain the mechanism for smell.

10.6 Sense of Taste

9. Explain the mechanism for taste.

10.7 Sense of Hearing

10. Explain the function of each part of the ear.

10.8 Sense of Equilibrium

11. Distinguish between static and dynamic equilibrium.

10.9 Sense of Sight

12. Explain the function of each part of the eye.
13. Explain how the eye refracts light.
14. Describe the visual nerve pathway.

FOCUS QUESTION

When you began this chapter at 3:00 P.M., it was 32°F outside, but sunlight was pouring into the room. It is now after 5:00 P.M., and, as you reach to turn on the light, you notice the room has become chilly, so you get a sweater. You smell the supper your roommate is preparing, and you realize you are hungry. How have your somatic and special senses functioned to process and act on this sensory information?

MASTERY TEST

Now take the mastery test. Do not guess. As soon as you complete the test, correct it. Note your successes and failures so that you can read the chapter to meet your learning needs.

1. List four kinds of receptors of general senses and what stimulates them. List receptors for special senses and what stimulates them.
2. Sensation of sensory impulses results from
   a. the type of receptor stimulated.
   b. the nature of impulse conduction.
   c. the region of the brain that receives the impulse.
   d. prior experience with the type of impulse.
3. The ability to ignore unimportant stimuli or become less responsive is called ______________________.
4. List the locations for receptors associated with the somatic senses.
5. Meissner's corpuscles and Pacinian corpuscles are sensitive to
   a. touch and pressure.
   b. pain.
   c. heat.
   d. light.
6. Sensory receptors for all of the following stimuli adapt to repeated stimulation by sending fewer and fewer impulses, *except* those for
   a. heat.
   b. light.
   c. pain.
   d. touch.
7. Which of the following events will elicit pain from visceral organs?
   a. spasm of smooth muscle
   b. cutting into the viscera
   c. stretching of a visceral organ
   d. burning, as in electrocautery
8. Sharp pain that disappears when the pain stimulus is stopped is conducted by ______________________.
9. In what area of the brain do most pain fibers terminate?
10. Pain-suppressing substances found in the pituitary gland and the hypothalamus are called ______________________.
11. List the special senses.
12. The receptors for taste and smell are examples of
    a. mechanical receptors.
    b. chemoreceptors.
    c. thermoreceptors.
13. Where are olfactory organs located?
14. Stimulated olfactory receptors transmit impulses to the ______________________.
15. The sensitive part(s) of a taste bud is (are) the
    a. taste cells.
    b. taste pore.
    c. taste hairs.
16. Saliva enhances the taste of food by
    a. increasing the motility of taste receptors.
    b. dissolving the chemicals that cause taste.
    c. releasing taste factors by partially digesting food.
17. The five primary taste sensations are ______________________, ______________________, ______________________, ______________________, and ______________________.
18. In addition to the sense of hearing, the ear also functions in the sense of ______________________.
19. The functions of the small bones of the middle ear are to
    a. provide a framework for the tympanic membrane.
    b. protect the structures of the inner ear.
    c. transmit vibrations from the external ear to the inner ear.
    d. increase the force of vibrations transmitted to the inner ear.
20. A means of providing equal pressure on both sides of the eardrum is furnished by the ______________________.
21. The inner ear consists of communicating chambers called a ______________________.
22. The inner ear is divided into three parts. The ______________________ and ______________________ provide a sense of equilibrium. The ______________________ is involved with hearing.

23. Hearing receptors are located in the
a. organ of Corti.
b. scala vestibuli.
c. scala tympani.
d. round window.

24. The hair cells of the vestibule are stimulated by
a. bending of the head forward or backward.
b. rapid turns of the head or body.
c. changes in the position of the body relative to the ground.
d. changes in the position of skeletal muscles.

25. The muscle that raises the eyelid is the
a. orbicularis oculi.
b. superior rectus.
c. levator palpebrae superioris.
d. ciliary muscle.

26. The conjunctiva covers the anterior surface of the eyeball, except for the ____________________________.

27. The superior rectus muscle rotates the eye
a. upward and toward the midline.
b. toward the midline.
c. away from the midline.
d. upward and away from the midline.

28. The transparency of the cornea is due to the
a. nature of the cytoplasm in the cells of the cornea.
b. small number of cells and the lack of blood vessels.
c. lack of nuclei within the cells of the cornea.
d. keratinization of cells in the cornea.

29. In the posterior wall of the eyeball, the sclera is pierced by the____________________________________.

30. The shape of the lens changes as the eye focuses on a close object in a process known as
a. accommodation.
b. refraction.
c. reflection.
d. strabismus.

31. The fluid that circulates through the anterior chamber and nourishes cells of cornea and lens is known as____________________________.

32. The part of the eye that controls the amount of light entering the eye is the ____________________________.

33. The inner tunic of the eye contains the receptor cells of sight and is called the ____________________________.

34. The region associated with the sharpest vision is the
a. macula lutea.
b. fovea centralis.
c. optic disk.
d. choroid coat.

35. The bending of light waves as they pass at an oblique angle from a medium of one optical density to a medium of another optical density is called ________________________.

36. There are two types of visual receptors: ________________ and ________________.

37. Match the type of vision in the first column with the proper receptor in the second column.

_____a. vision in relatively dim light
_____b. color vision
_____c. general outlines
_____d. sharp images

1. rods
2. cones

38. The light-sensitive pigment in rods is ______________________. In the presence of light, this pigment decomposes to form __________________________ and __________________________.

39. Some of the fibers of the optic nerves cross within the ________________________________.

STUDY ACTIVITIES

Aids to Understanding Words

Define the following words and word parts. (p. 263)

choroid

cochlea

iris

labyrinth

lacri-

macula

olfact-

scler-

tympan-

vitre-

10.1 Introduction (p. 263)

A. What is the function of sensory receptors?

B. List the general senses and the special senses. How are they different?

10.2 Receptors, Sensations, and Perception (p. 263)

A. List five groups of sensory receptors.

B. The process that allows an individual to locate the region of stimulation is called ________________________.

C. The process that makes a receptor ignore a continuous stimulus unless the strength of that stimulus is increased is ______________________________.

D. What is a sensation? What is a perception?

10.3 General Senses (pp. 264–267)

A. Fill in the following table regarding the general senses.

Type	Structure/location	Sensation
Free nerve endings		
Meissner's corpuscles		
Pacinian corpuscles		
Temperature senses		
Free nerve endings		

B. Answer these questions concerning pain receptors. (pp. 265–267)

1. How do pain receptors differ from the other somatic senses?

2. What events trigger visceral pain?

3. What is referred pain?

4. Compare acute pain fibers and chronic pain fibers.

5. How does the brain regulate pain impulses?

10.4 Special Senses (p. 267)

List the special senses.

10.5 Sense of Smell (pp. 267–269)

A. The sense of smell supplements the sense of ______________________________. What type of receptor is involved in these sensations? (p. 267)

B. On the accompanying illustrations (p. 268), label the olfactory tract, olfactory bulb, cribriform plate, nasal cavity, olfactory area of the nasal cavity, superior nasal concha, nerve fibers within olfactory bulb, cilia, olfactory receptor cells, and columnar epithelial cells.

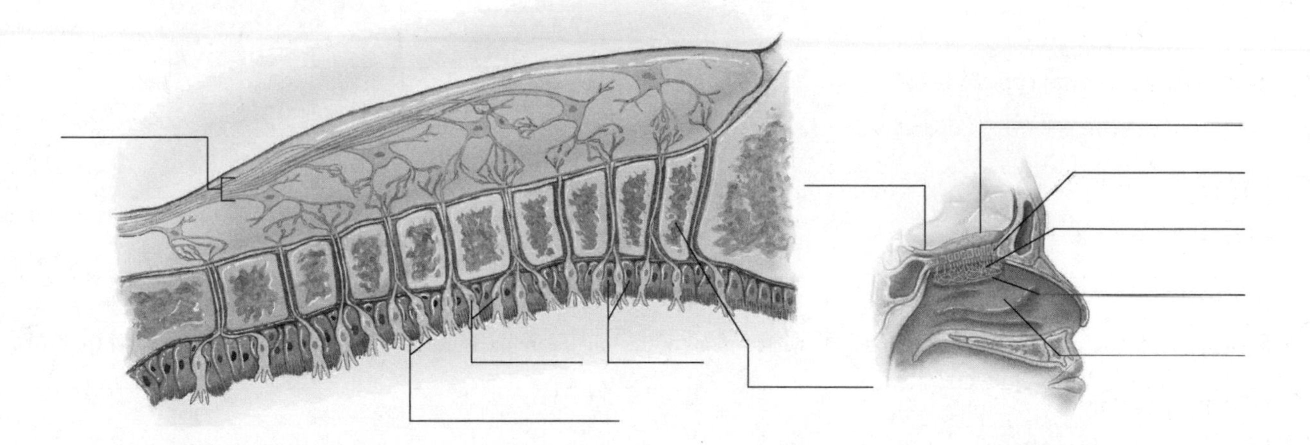

C. How do odors cause olfactory stimulation? (pp. 268–269)

10.6 Sense of Taste (pp. 269–270)

A. Describe the structure of taste receptors. (p. 269)

B. How does saliva contribute to the perception of taste? (p. 270)

C. Label the following structures in the accompanying illustration of taste buds (p. 269): papillae, taste bud, epithelium of the tongue, taste cell, taste hair, supporting cell, taste pore, sensory nerve fibers, connective tissue.

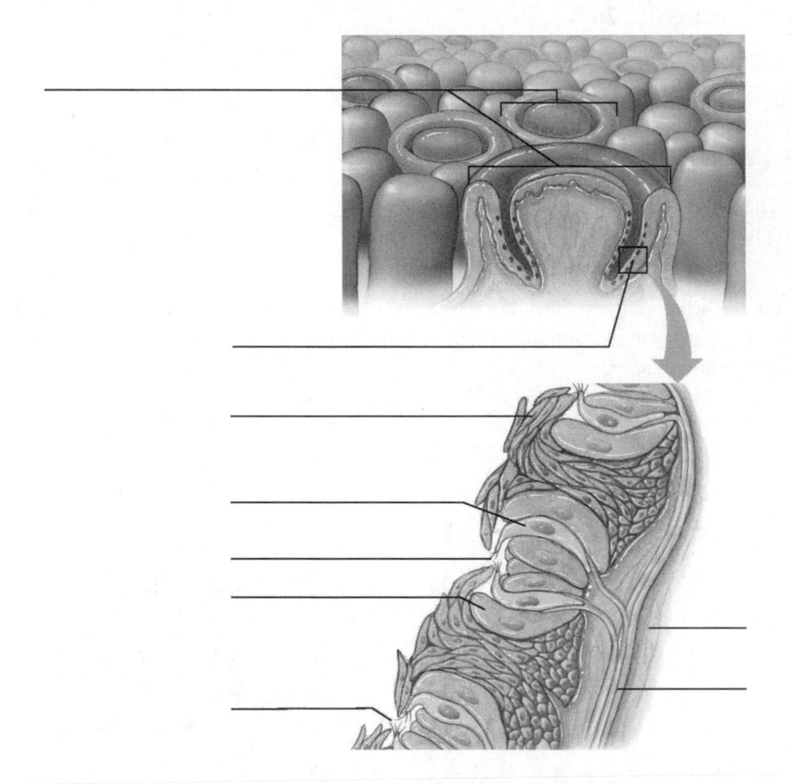

D. List five primary taste sensations. On what regions of the tongue are the sensations for each taste the strongest? (p. 270)

E. Describe the nerve pathways for taste. (p. 270)

10.7 Sense of Hearing (pp. 270–275)

A. Describe the function of the external ear. (p. 270)

B. Describe the structure and function of the middle ear. (p. 271)

C. What is the purpose of the auditory tube and why does it help to chew gum while descending in an airplane? (pp. 271–272)

D. Describe the function of the inner ear. (pp. 272–274)

E. Describe the structures and pathway involved in vibration and generation of sensation from the auditory meatus to the temporal lobe of the cerebrum. (pp. 274–275)

F. Explain the difference between conductive and sensorineural deafness. (p. 275)

10.8 Sense of Equilibrium (pp. 275–276)

A. Distinguish between static and dynamic equilibrium. (p. 275)

B. Describe the function of each of the following structures in maintaining equilibrium and distinguish if the role is in static, dynamic, or both types of equilibrium.

utricle

saccule

macula

semicircular canals

crista ampullaris

cerebellum

eyes

10.9 Sense of Sight (pp. 276–287)

A. Answer these questions concerning the visual accessory organs. (pp. 276–279)

1. What structures are covered by the conjunctiva?

2. Describe the lacrimal apparatus. How does it protect the eye?

3. Identify the function of the following muscles: orbicularis oculi, medial rectus, levator palpebrae superioris, lateral rectus, superior rectus, superior oblique, inferior rectus, inferior oblique.

B. Label the following structures in the accompanying illustration and assign each of the structures a function (p. 280): cornea, lens, iris, suspensory ligaments, vitreous humor, aqueous humor, sclera, optic disk, optic nerve, fovea centralis, retina, choroid coat, pupil, ciliary body, anterior cavity, posterior cavity, anterior chamber, posterior chamber, lateral rectus, medial rectus.

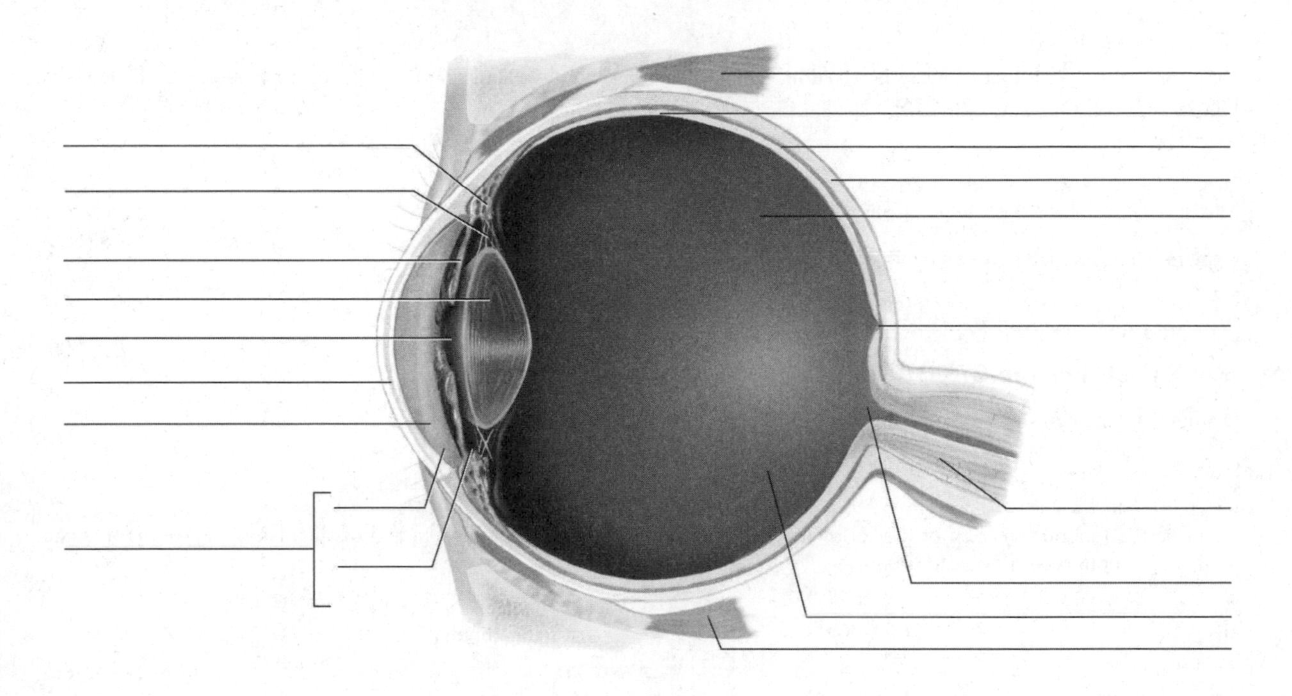

C. Describe what accommodation is and the structures involved in the process. (pp. 280–281)

D. What is refraction and how do structures of the eye influence refraction of light? (pp. 284–285)

E. Answer these questions concerning photoreceptors. (pp. 285–286)

1. Describe the functions of rods.

2. Describe the location and functions of cones.

3. How do cones distinguish color?

F. Describe the visual nerve pathway. (pp. 286–287)

Clinical Focus Questions

A. Compare the loss of vision due to cataracts and glaucoma. How is each of these conditions treated?

B. What are the symptoms of cataracts? of glaucoma?

C. How many cranial nerves are devoted to the sense of sight?

D. What conclusions can be drawn based on your response to question C?

When you have finished the study activities to your satisfaction, retake the mastery test and compare your results with your initial attempt. If you are not satisfied with your performance, repeat the appropriate study activities.

CHAPTER 11

ENDOCRINE SYSTEM

OVERVIEW

The endocrine system, like the nervous system, controls body activities to maintain a relatively constant internal environment. The methods used by these two systems are different. This chapter defines hormones and describes the different methods of secretion (learning outcomes 1 and 2). The difference between endocrine and exocrine glands is defined, and the location of the endocrine glands and the hormones they secrete are discussed (learning outcomes 3–5). It explains the general function of hormones, how hormones affect target tissues, how the secretion of hormones is controlled by a negative feedback system and the nervous system, and the result of too little or too much of each hormone (learning outcomes 6–12). Lastly, the mediation of stress is discussed (learning outcome 13).

Knowledge of the function of the endocrine system is basic to understanding how metabolic processes are regulated to meet the changing needs of the body.

LEARNING OUTCOMES

After you have studied this chapter, you should be able to

11.1 Introduction

1. Describe the secretions of the endocrine system.
2. Distinguish between paracrine and autocrine secretions.
3. Distinguish between endocrine and exocrine glands.

11.2 General Characteristics of the Endocrine System

4. Explain how the nervous and endocrine systems are alike and how they are different.
5. Describe the source of specificity of the endocrine system.
6. Name some functions of hormones.

11.3 Hormone Action

7. Explain how steroid and nonsteroid hormones affect target cells.

11.4 Control of Hormonal Secretions

8. Discuss how negative feedback mechanisms regulate hormonal secretions.
9. Explain how the nervous system controls secretion.

11.5–11.10 Pituitary Gland–Other Endocrine Glands

10. Name and describe the locations of the major endocrine glands, and list the hormones they secrete.
11. Describe the functions of the hormones that endocrine glands secrete.
12. Explain how the secretion of each hormone is regulated.

11.11 Stress and Health

13. Describe how the body responds to stress.

FOCUS QUESTION

How does the endocrine system complement the nervous system in maintaining a person's ability to interact with both external and internal environments?

MASTERY TEST

Now take the mastery test. Do not guess. As soon as you complete the test, correct it. Note your successes and failures so that you can read the chapter to meet your learning needs.

1. Chemical signals sent between individuals are known as ______________________________.

2. A chemical messenger involved in the regulation of body functions is a ______________.

3. "Local" hormones that affect only neighboring cells are ______________ hormones; hormones secreted by a cell that affect only the secreting cell are ______________ hormones.

4. Glands that secrete substances outside the internal environment are ______________________ glands.

5. Glands that release their secretions directly into the body fluids and help regulate metabolic processes are ______________________ glands.

6. Many hormones are thought to function by acting on receptor sites in the ____________________.

7. A major difference between steroid and nonsteroid hormones is the degree to which they
   a. can be manufactured as drugs.
   b. are lipid-soluble.
   c. can be excreted by the kidney.
   d. require metabolism in the liver.

8. Another group of compounds that have hormonelike effects and are synthesized from arachidonic acid are the ______________________________.

9. The characteristics of the negative feedback system that regulate hormone secretion include
   a. activation by an imbalance.
   b. exertion of an inhibitory effect on the gland.
   c. exertion of a stimulating effect on the gland.
   d. a tendency for levels of hormone to fluctuate wildly.

10. The region of the brain most responsible for controlling endocrine function is the ______________________.

11. The anterior and posterior lobes of the pituitary gland are regulated by the ______________________.

12. The hormones secreted by the anterior lobe of the pituitary gland include
    a. thyroid-stimulating hormone.
    b. luteinizing hormone.
    c. antidiuretic hormone.
    d. oxytocin.

13. Which of the following are actions of pituitary growth hormone?
    a. enhance the movement of amino acids through the cell membrane
    b. increase the utilization of glucose by cells
    c. increase the utilization of fats by cells
    d. enhance the movement of potassium across the cell membrane

14. The pituitary hormone that stimulates and maintains milk production following childbirth is ______________.

15. TSH secretion is regulated by
    a. circulating thyroid hormones.
    b. blood sugar levels.
    c. osmolarity of blood.
    d. TRH secreted by the hypothalamus.

16. Which of the following pituitary hormones regulate(s) reproductive function?
    a. LSH
    b. ADH
    c. FSH
    d. oxytocin

17. A hormone of the posterior pituitary gland plays a major role in regulating the amount of water in the body.
    a. True
    b. False

18. The thyroid hormones that affect the metabolic rate are ______________________ and ______________________.

19. Which of the following are functions of thyroid hormones?
    a. control sodium levels
    b. decrease rate of energy release from carbohydrates
    c. increase protein synthesis
    d. normal growth in children

20. The element necessary for normal function of the thyroid gland is ______________________.

21. The hormone produced by the thyroid gland responsible for blood calcium regulation is ______________________.

22. Hypothyroidism in an infant is characterized by
    a. hyperactivity.
    b. mental retardation.
    c. excessive appetite.
    d. abnormal bone formation.

23. Which of the following statements about parathyroid hormone is (are) true?
a. Parathyroid hormone enhances absorption of calcium from the intestine.
b. Parathyroid hormone stimulates bone to release ionized calcium.
c. Parathyroid hormone stimulates the kidneys to conserve calcium.
d. Parathyroid hormone secretion is stimulated by the hypothalamus.

24. Injury to or removal of parathyroid glands is likely to result in
a. reduced osteoclastic activity.
b. Cushing's disease.
c. kidney stones.
d. hypercalcemia.

25. The hormones of the adrenal medulla are ______________________________ and ______________________________.

26. The adrenal hormone aldosterone belongs to a category of cortical hormones called
a. mineralocorticoids.
b. glucocorticoids.
c. sex hormones.

27. The actions of cortisol include
a. increase in protein synthesis to decrease the levels of circulating amino acids.
b. increased release of fatty acids and decreased use of glucose.
c. stimulation of gluconeogenesis.
d. conservation of water.

28. Adrenal sex hormones are primarily (male, female).

29. The endocrine portion of the pancreas consists of groups of cells called ____________________________.

30. The hormone that responds to low blood sugar by stimulating the liver to convert glycogen to glucose is ____________________________.

31. The action(s) of insulin that most directly lead(s) to lowering blood sugar levels is (are)
a. enhancing glucose absorption from the small intestine.
b. facilitating the transport of glucose across the cell membrane.
c. promoting the transport of amino acids out of the cell.
d. increasing the synthesis of fats.

32. The most common type of diabetes mellitus is (type I, type II).

33. The hormone melatonin is secreted by the
a. thymus.
b. pineal gland.
c. gonads.

34. A negative side effect of the general stress syndrome is increased susceptibility to ______________________________.

35. Stressors may be positive or negative stimuli.
a. True
b. False

36. What gland activates the fight-or-flight mechanism?
a. hypothalamus
b. pituitary
c. pancreas
d. adrenal medulla

STUDY ACTIVITIES

Aids to Understanding Words

Define the following word parts. (p. 292)

-crin

diuret-

endo-

exo-

hyper-

hypo-

para-

toc-

-tropic

11.1 Introduction (p. 292)

A. What is a hormone?

B. List the differences between endocrine and exocrine glands.

C. What are paracrine and autocrine secretions?

11.2 General Characteristics of the Endocrine System (pp. 292–293)

A. As a group, endocrine glands regulate ________________________________.

B. List the ways in which endocrine glands accomplish their general function. (pp. 292–293)

C. List the major endocrine glands. (p. 293)

11.3 Hormone Action (pp. 293–296)

A. What are the two types of hormones? (p. 293)

B. Describe how hormones are synthesized. (p. 293)

C. Compare the characteristics of steroid and nonsteroid hormones. (pp. 294–295)

D. What is the role of cyclic adenosine monophosphate? (p. 295)

E. What are prostaglandins? (p. 296)

11.4 Control of Hormonal Secretions (pp. 296–297)

A. List the three mechanisms that control hormone secretion. (pp. 296–297)

B. Describe how the negative feedback system regulates hormone secretion. (p. 296–297)

11.5 Pituitary Gland (pp. 297–301)

A. Describe the structure and location of the pituitary gland. (p. 297)

B. Fill in the following table regarding pituitary hormones. (pp. 298–301)

Hormone	Source of control	Actions
Anterior lobe		
Growth hormone		
Prolactin		
Thyroid-stimulating hormone		
Adrenocorticotropic hormone		
Follicle-stimulating hormone		
Luteinizing hormone		
Posterior lobe		
Antidiuretic hormone		
Oxytocin		

11.6 Thyroid Gland (pp. 301–303)

A. In the accompanying illustration (p. 302), label the thyroid gland, isthmus, larynx, colloid, follicular cell, and extrafollicular cell.

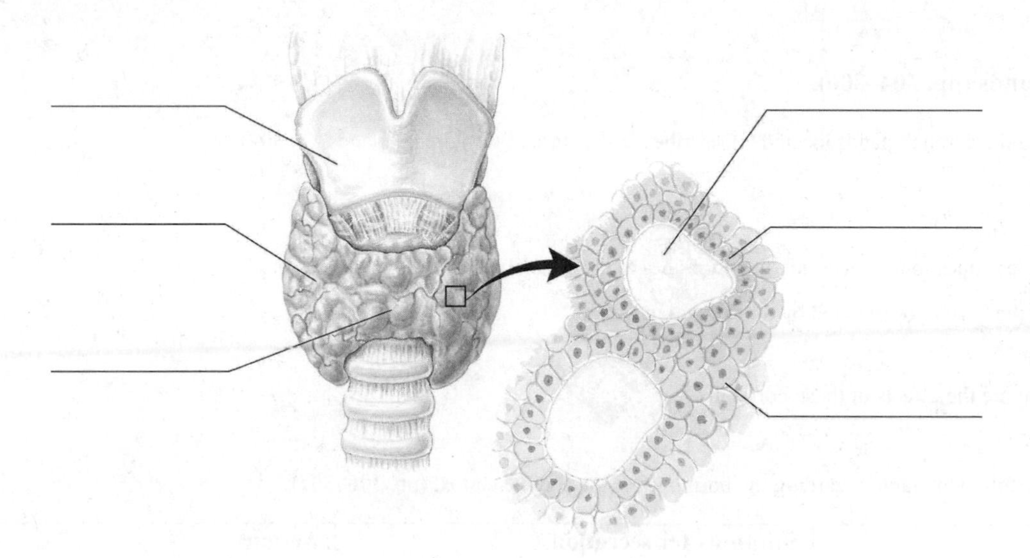

B. Answer these questions concerning thyroid hormones and their functions. (pp. 301–302)

1. What element is needed to synthesize thyroxine and triiodothyronine?

2. What are the functions of thyroxine and triiodothyronine?

3. What is the role of calcitonin?

4. What cells synthesize calcitonin and how is it regulated?

11.7 Parathyroid Glands (pp. 303–304)

A. Where are the parathyroid glands located? (p. 303)

B. Describe how parathyroid hormone affects blood levels of calcium and phosphorus. Include its effect on bone, the intestine, and the kidneys. (p. 303)

11.8 Adrenal Glands (pp. 304–306)

A. Where are the adrenal glands located? Describe the structure of the adrenal glands. (p. 304)

B. Answer these questions concerning the hormones of the adrenal medulla. (pp. 304–305)

1. List the hormones secreted by the adrenal medulla.

2. What are the effects of these hormones?

C. Fill in the following table regarding the hormones of the adrenal cortex. (pp. 306–307)

Hormone	Stimulus for secretion	Action
Mineralocorticoids (e.g., aldosterone)		
Glucocorticoid (e.g., cortisol)		
Sex hormones (androgens)		

D. Describe the negative feedback mechanism that controls the release of cortisol. (p. 306)

11.9 Pancreas (pp. 306–309)

A. Describe the structure and location of the pancreas. (p. 307)

B. Fill in the following table regarding hormones of the pancreas. (pp. 308–309)

Hormone	Stimulus for secretion	Action
Glucagon		
Insulin		

11.10 Other Endocrine Glands (pp. 309–310)

A. Where is the pineal gland located, and what is its function? (p. 309)

B. Where is the thymus gland located, and what is its function? (pp. 309–310)

C. List the names and sources of the reproductive hormones. (p. 310)

11.11 Stress and Health (pp. 311–313)

A. List examples of events that may qualify as stressors to the body. (p. 311)

B. Describe the stress response or general adaptation syndrome. (pp. 311–312)

C. Describe the alarm stage and address how it alleviates stress. (pp. 311–312)

D. Describe the resistance stage and address why it has negative effects on a person's health. (p. 312)

Clinical Focus Question

How do food intake, exercise, and insulin interact to control levels of glucose in the blood? How does stress affect this process?

When you have finished the study activities to your satisfaction, retake the mastery test and compare your results with your initial attempt. If you are not satisfied with your performance, repeat the appropriate study activities.

UNIT 4

TRANSPORT

CHAPTER 12

BLOOD

OVERVIEW

This chapter enables you to describe the general characteristics of blood and discuss its major functions (learning outcome 1). Your study of blood cells allows you to explain the significance of blood cell counts in assessing the health of an individual, describe how the production of red blood cells is controlled, and describe the characteristics of the five types of white blood cells and the function of each type (learning outcomes 3–5). Upon achieving learning outcomes 2 and 6, you will be able to describe the various components of the cellular matrix of blood and plasma, as well as describe the functions of each component. Hemostasis and the nature of blood typing and transfusion, as well as the potential reaction between fetal and maternal blood tissues, are explained in learning outcomes 7–10.

LEARNING OUTCOMES

After you have studied this chapter, you should be able to

12.1 Introduction

1. Describe the general characteristics of blood, and discuss its major functions.
2. Distinguish among the formed elements and liquid portion of blood.

12.2 Blood Cells

3. Explain the significance of red blood cell counts.
4. Summarize the control of red blood cell production.
5. Distinguish among the five types of white blood cells, and give the function(s) of each type.

12.3 Blood Plasma

6. Describe the functions of each of the major components of plasma.

12.4 Hemostasis

7. Define *hemostasis* and explain the mechanisms that help achieve it.
8. Review the major steps in blood coagulation.

12.5 Blood Groups and Transfusions

9. Explain blood typing and how it is used to avoid adverse reactions following blood transfusions.
10. Describe how blood reactions may occur between fetal and maternal tissues.

FOCUS QUESTION

How does the structure of the components of blood help meet oxygenation needs, allow recognition and rejection of foreign protein, and control coagulation of the blood?

MASTERY TEST

Now take the mastery test. Do not guess. As soon as you complete the test, correct it. Note your successes and failures so that you can read the chapter to meet your learning needs.

1. Blood is considered a type of ______________________________ tissue.
2. Plasma represents ______________________________ % of a normal blood sample.
3. Which of the following formed elements of blood are essential to gas exchange?
   a. platelets
   b. red blood cells
   c. white blood cells
4. Which of the following statements is (are) true?
   a. Red blood cells have no nucleus.
   b. Red blood cells lose their nuclei to make more room for hemoglobin.
   c. Red blood cells use none of the oxygen they carry.
   d. Red blood cells have a limited life span because they cannot synthesize protein.
5. The normal red blood cell count for an adult male is ______________________________ per microliter; and it is ______________________________ per microliter for an adult female.
6. Increases in the number of red blood cells increase the oxygen-carrying capacity of the blood.
   a. True
   b. False
7. What is the average life span of a red blood cell?
8. Red blood cell production is stimulated by the hormone ______________________________, which is released from the kidney in response to low oxygen concentration.
9. After birth, red blood cells are produced in the
   a. spleen.
   b. red marrow.
   c. yellow marrow.
   d. liver.
10. Which of the following nutrients is (are) required for optimum red blood cell production?
    a. vitamin B_{12}
    b. calcium
    c. folic acid
    d. iodine
11. What mineral is necessary for the production of normal red blood cells?
12. The heme portion of damaged red blood cells is decomposed into iron and
    a. biliverdin.
    b. bilirubin.
    c. bile.
    d. granulocytes.
13. Damaged red blood cells are destroyed by large phagocytic cells called
    a. lymphocytes.
    b. macrophages.
    c. neutrophils.
    d. granulocytes.
14. The cytokines that stimulate white blood cell production are ______________________________ and ______________________________.
15. The most numerous type of white blood cell is the
    a. neutrophil.
    b. eosinophil.
    c. monocyte.
    d. lymphocyte.
16. The white blood cell that forms antibodies necessary for immunity to specific diseases is the
    a. basophil.
    b. lymphocyte.
    c. thrombocyte.
    d. eosinophil.
17. The normal white blood cell count is ______________________________ to ______________________________ per microliter of blood.
18. White blood cells combat infection by
    a. phagocytosis.
    b. pinocytosis.
    c. production of antibodies.
    d. production of enzymes.
19. The substances found in the cytoplasmic granules of basophils include ______________________________ and ______________________________.
20. The hormone responsible for the production of platelets is ______________________________.

21. Match the functions and characteristics in the first column with the appropriate plasma proteins in the second column.

_____a. most plentiful protein(s) in plasma
_____b. significant in maintaining osmotic pressure
_____c. transport(s) lipids and fat-soluble vitamins
_____d. antibody(ies) of immunity
_____e. play(s) a part in blood clotting

1. albumins
2. globulins
3. fibrinogen

22. List the gases found in blood plasma and note which of these gases has no physiological function.

23. List the nutrients found in plasma.

24. The most abundant plasma electrolytes are
a. calcium.
b. sodium.
c. potassium.
d. chlorides.

25. A platelet plug begins to form when platelets are
a. exposed to air.
b. exposed to a rough surface.
c. exposed to calcium.
d. crushed.

26. The major event in the formation of a blood clot is the transformation of a soluble plasma protein, ___________________________, to a relatively insoluble protein, ______________________________.

27. Prothrombin is a plasma protein that is produced by the
a. kidney.
b. small intestine.
c. pancreas.
d. liver.

28. Once a blood clot begins to form, it promotes still more clotting. This is an example of a(n) ___________________________ system.

29. A fragment of a blood clot that is traveling in the bloodstream is called a(n) ___________________________.

30. The clumping together of red blood cells when different types of blood are mixed is due to antibodies in the plasma and antigens on the
a. thrombocytes.
b. erythrocytes.
c. basophils.
d. eosinophils.

31. A person with type A blood has
a. agglutinogen A and agglutinin B.
b. agglutinogens A and B.
c. agglutinins A and B.
d. neither agglutinin A nor B.

32. Agglutinins for Rh appear
a. spontaneously as an inherited trait.
b. only rarely for poorly understood reasons.
c. only in response to stimulation by Rh agglutinogens.

STUDY ACTIVITIES

Aids to Understanding Words

Define the following word parts. (p. 319)

agglutin-

bil-

embol-

erythr-

hema-

hemo-

leuko-

-osis

-poie

-stasis

thromb-

12.1 Introduction (p. 319)

A. Describe blood and list the components.

B. What is the function of blood?

C. What is the volume of blood and what factors influence this?

12.2 Blood Cells (pp. 319–327)

A. Answer these questions concerning red blood cells. (pp. 319–321)

1. The shape of a red blood cell is a(n) ____.
2. How does the shape enhance the function of red blood cells?
3. Red blood cells are ________________ red when carrying oxygen and are ______________red when oxygen is released.
4. What organelles do mature red cells lack and how does this influence their function?

B. Answer these questions concerning red blood cell counts and production. (pp. 321–323)

1. What is the normal red blood cell count for a man? What about a woman?
2. Where are red blood cells produced?
3. How is the production of red blood cells controlled?
4. What factors would cause or contribute to a normal increase in red blood cells?
5. What dietary factors influence red blood cell production?
6. How does sickle cell disease affect red blood cell production?

C. Answer these questions concerning the destruction of red blood cells. (pp. 323–324)

1. How are red blood cells damaged?

2. Damaged red blood cells are destroyed by cells called ______________________. Destruction of red blood cells primarily occurs in the ____________________ and __________________.

3. What happens to the liberated hemoglobin?

D. Fill in the following table regarding the different types of white blood cells. (pp. 324–326)

White blood cell	**Description**	**Percentage of total**	**Function**
Granulocytes			
Neutrophil			
Eosinophil			
Basophil			
Agranulocytes			
Lymphocyte			
Monocyte			

E. Answer these questions concerning white blood cell counts. (pp. 325–327)

1. What is a normal white blood cell count?

2. What causes an increase or a decrease in white blood cells?

3. What is a differential white blood cell count?

F. Describe the structure and function of platelets. (p. 327)

12.3 Blood Plasma (pp. 327–330)

A. Fill in the following table regarding the plasma proteins. (p. 329)

Protein	Description	Percentage of total	Function
Albumins			
Globulins			
Fibrinogens			

B. Answer these questions concerning nutrients and gases. (pp. 329–330)

1. What gases are found in plasma?

2. What nutrients are found in plasma?

3. What are the nonprotein nitrogenous substances found in plasma and from where do they come?

C. Answer these questions concerning plasma electrolytes. (p. 330)

1. What electrolytes are found in plasma?

2. What are the sources of the electrolytes and which ones are the most abundant?

12.4 Hemostasis (pp. 330–333)

A. What is hemostasis and what are the three mechanisms involved in the process? (p. 330)

B. How is a platelet plug formed? (p. 330)

C. Describe the major events that lead to blood coagulation. (pp. 331–332)

D. Describe the mechanisms that limit or prevent coagulation. (pp. 331–332)

E. What is a thrombus? What is an embolus? What conditions predispose a person to the formation of thrombi? (pp. 331–332)

12.5 Blood Groups and Transfusions (pp. 333–336)

A. Answer these questions concerning agglutinogens and agglutinins. (p. 333)

1. What are antigens (agglutinogens) and antibodies (agglutinins)?

2. Agglutinogens are present on the ____; agglutinins are in the ____________________________.

B. Answer these questions concerning the ABO blood group. (pp. 334–336)

1. Describe the basis for ABO blood types.

2. Why is it unsafe to mix different blood types?

C. Answer these questions concerning the Rh blood group. (p. 336)

1. What is the Rh blood group?

2. How does Rh antibody formation differ from A, B, and O agglutinin formation?

3. How does a fetus develop erythroblastosis fetalis?

Clinical Focus Questions

Fetal hemoglobin is constructed to maximize oxygen-carrying capacity *in utero*. This is important in understanding the development of symptoms of sickle cell disease.

A. Based on your knowledge of the life cycle of red blood cells, when do you expect the symptoms of sickle cell disease to appear?

B. What is the pathologic event that leads to sickle cell disease?

C. Why are children with sickle cell disease susceptible to infection?

D. What recommendations would you make to minimize the damage of sickle cell disease?

When you have finished the study activities to your satisfaction, retake the mastery test and compare your results with your initial attempt. If you are not satisfied with your performance, repeat the appropriate study activities.

CHAPTER 13

CARDIOVASCULAR SYSTEM

OVERVIEW

This chapter deals with the system that transports blood to and from cells—the cardiovascular system. It identifies the major organs of the cardiovascular system and explains their functions (learning outcome 1). It discusses the location, structure, and function of the parts of the heart and the major types of blood vessels (learning outcomes 2, 3, and 8) and explains the pulmonary, systemic, and coronary circuits of the cardiovascular system and the major vessels in each circuit (learning outcomes 4 and 12). It discusses the cardiac cycle and its control and relates these to the normal ECG (learning outcomes 5–7). In addition, it describes how blood pressure is created and controlled, how venous blood is returned to the heart, and how substances are exchanged between capillary blood and tissue fluid (learning outcomes 9–11).

After learning about the blood, study of the cardiovascular system demonstrates how oxygen is transported to cells and how waste products are transported away from cells to maintain function. Finally, the major arteries and veins of the circulatory system are identified and located (learning outcome 13).

LEARNING OUTCOMES

After you have studied this chapter, you should be able to

13.1 Introduction

1. Name the structures composing the cardiovascular system.

13.2 Structure of the Heart

2. Distinguish between the coverings of the heart and the layers that compose the wall of the heart.
3. Identify and locate the major parts of the heart, and discuss the functions of each part.
4. Trace the pathway of blood through the heart and the vessels of coronary circulation.

13.3 Heart Actions

5. Describe the cardiac cycle and the cardiac conduction system.
6. Identify the parts of a normal ECG pattern, and discuss the significance of this pattern.
7. Explain control of the cardiac cycle.

13.4 Blood Vessels

8. Compare the structures and functions of the major types of blood vessels.
9. Describe how substances are exchanged between blood in capillaries and the tissue fluid surrounding body cells.

13.5 Blood Pressure

10. Explain how blood pressure is produced and controlled.
11. Describe the mechanisms that aid in returning venous blood to the heart.

13.6 Paths of Circulation

12. Compare the pulmonary and systemic circuits of the cardiovascular system.

13.7–13.8 Arterial System–Venous System

13. Identify and locate the major arteries and veins.

FOCUS QUESTION

How does the cardiovascular system maintain blood flow and pressure when walking up the steps to your class and when sitting in class taking notes?

MASTERY TEST

Now take the mastery test. Do not guess. As soon as you complete the test, correct it. Note your successes and failures so that you can read the chapter to meet your learning needs.

1. The heart is a cone-shaped, muscular pump located within the ________________________.

2. The base of the heart is located
   a. behind the second rib.
   b. under the sternum.
   c. between the fourth and fifth rib.

3. The visceral pericardium is also known as the
   a. epicardium.
   b. myocardium.
   c. endocardium.

4. Inflammation of a membrane covering the heart is known as ____________________________.

5. List the layers of the wall of the heart.

6. Purkinje fibers are located in the
   a. epicardium.
   b. myocardium.
   c. endocardium.
   d. parietal pericardium.

7. The upper chambers of the heart are the right and left ________________________; the lower chambers are the right and left ___________________________.

8. The vessels that empty into the upper right chamber of the heart are
   a. inferior and superior venae cavae.
   b. pulmonary veins.
   c. pulmonary arteries.
   d. coronary sinuses.

9. The valve between the chambers of the left side of the heart is the
   a. aortic semilunar valve.
   b. bicuspid valve (mitral valve).
   c. tricuspid valve.
   d. pulmonary semilunar valve.

10. Strong fibrous strings attached to the cusps of the tricuspid and bicuspid valves and the papillary muscle are the ___________________________________.

11. Blood is supplied to the heart by the right and left_______________________________________.

12. Atrial contraction, while the ventricles are relaxed, followed by ventricular contraction, while the atria are relaxed, is known as the _________________________________.

13. Heart sounds are a result of
    a. blood entering the atria in large volumes.
    b. contraction of the myocardium.
    c. opening and closing of heart valves.
    d. changes in the blood flow rate through the chambers of the heart.

14. A mass of merging cells that function as a unit is called
    a. smooth muscle.
    b. functional syncytium.
    c. the sinoatrial node.
    d. the cardiac conduction system.

15. The cells that initiate the stimulus for contraction of the heart muscle are located in the
    a. sinoatrial node.
    b. atrioventricular node.
    c. Purkinje fibers.
    d. bundle of His.

16. A recording of the electrical changes that occur in the myocardium during the cardiac cycle is a(n) _______________.

17. In the recording described in question 16, atrial depolarization is represented by the
    a. P wave.
    b. QRS complex.
    c. T wave.
    d. U wave.

18. The effect of an increase of parasympathetic nerve impulses on the heart is to (decrease, increase) the heart rate.

19. Abnormalities in the concentration of which of the following ions is likely to interfere with contraction of the heart?
a. chloride
b. potassium
c. calcium
d. sodium

20. When the smooth muscle of the artery contracts, the action is called ____________________.

21. Fatty materials, particularly cholesterol, form deposits called ____________________ on the inner walls of arteries when the condition ____________________ occurs.

22. The vessel that participates directly in the exchange of substances between the cell and the blood is the
a. arteriole.
b. artery.
c. capillary.
d. venule.

23. The amount of blood that flows into capillaries is regulated by
a. constriction and dilation of capillaries.
b. arterioles.
c. the amount of intercellular tissue.
d. precapillary sphincters.

24. The transport mechanisms used by the capillaries are ____________________, ____________________, and ____________________.

25. Blood pressure is highest in
a. an artery.
b. an arteriole.
c. a capillary.
d. a vein.

26. Plasma proteins help retain water in the blood by maintaining
a. osmotic pressure.
b. hydrostatic pressure.
c. a vacuum.

27. The middle layer of the walls of veins differs from that of the arteries in that
a. it contains more connective tissue.
b. it contains less smooth muscle.
c. this layer is thicker in the vein.
d. it contains some striated muscle.

28. Blood in veins is kept flowing in one direction by the presence of ____________________.

29. The maximum pressure in the artery, occurring during ventricular contraction, is
a. diastolic pressure.
b. systolic pressure.
c. mean arterial pressure.
d. pulse pressure.

30. The amount of blood pushed out of the ventricle with each contraction is called ____________________.

31. List the four factors that influence blood pressure.

32. Starling's law is related to which of the following cardiac structures?
a. interventricular septum
b. conduction system
c. muscle fibers
d. heart valves

33. When the baroreceptors in the aorta and carotid artery sense an increase in blood pressure, the medulla relays (sympathetic, parasympathetic) impulses.

34. Peripheral resistance is maintained by increasing or decreasing the size of
a. capillaries.
b. arterioles.
c. venules.

35. Venous blood flow is maintained by all but which of the following factors?
a. blood pressure
b. skeletal muscle contraction
c. vasoconstriction of veins
d. respiratory movements

36. Which of the following vessels carries deoxygenated blood?
a. aorta
b. innominate artery
c. basilar artery
d. pulmonary artery

37. The pulmonary veins deliver blood to the ____________________.

38. List the arteries that originate from the arch of the aorta.

39. The abdominal aorta ends with the right and left ____________________ arteries.

STUDY ACTIVITIES

Aids to Understanding Words

Define the following word parts. (p. 341)

brady-

diastol-

-gram

papill-

syn-

systol-

tachy-

13.1 Introduction (p. 341)

A. What is the function of the cardiovascular system?

B. What are the two divisions of the cardiovascular system?

13.2 Structure of the Heart (pp. 342–347)

A. Describe the precise location of the heart. (p. 342)

B. Answer these questions concerning the coverings of the heart. (pp. 342–343)

1. The heart is enclosed by a double-layered ________________________________.
2. What is the function of the fluid in the pericardial space?
3. Describe the pathologic events of pericarditis.

C. Answer these questions concerning the wall of the heart. (p. 343)

1. Label these structures in the accompanying illustration: epicardium, myocardium, endocardium, fibrous pericardium, parietal pericardium, pericardial cavity, coronary blood vessel.

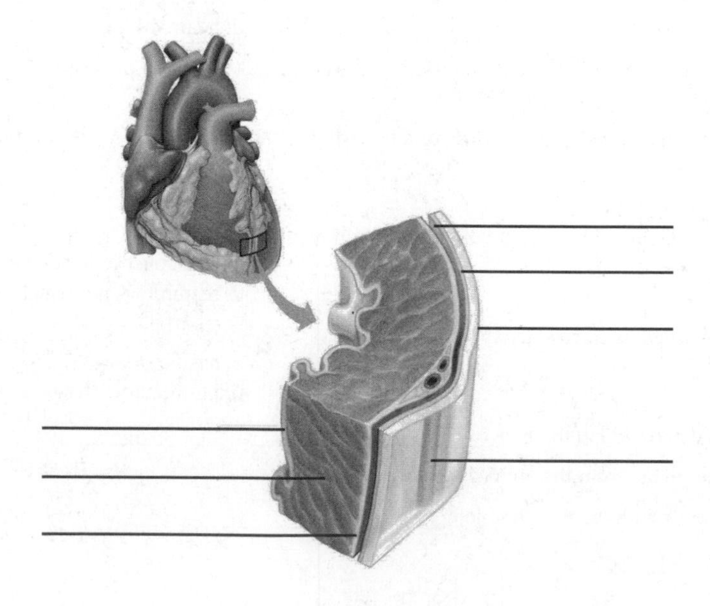

2. Describe the function of each labeled portion of the wall of the heart.

D. Answer these questions concerning heart chambers and valves. (pp. 343–345)

1. List the chambers of the heart.

2. Label these structures in the accompanying illustration: right and left ventricles, right and left atria, superior and inferior venae cavae, aorta, tricuspid valve, pulmonary valve, aortic valve, bicuspid valve, right and left pulmonary veins, right and left pulmonary arteries, chordae tendineae, papillary muscles, interventricular septum, pulmonary trunk, opening of coronary sinus.

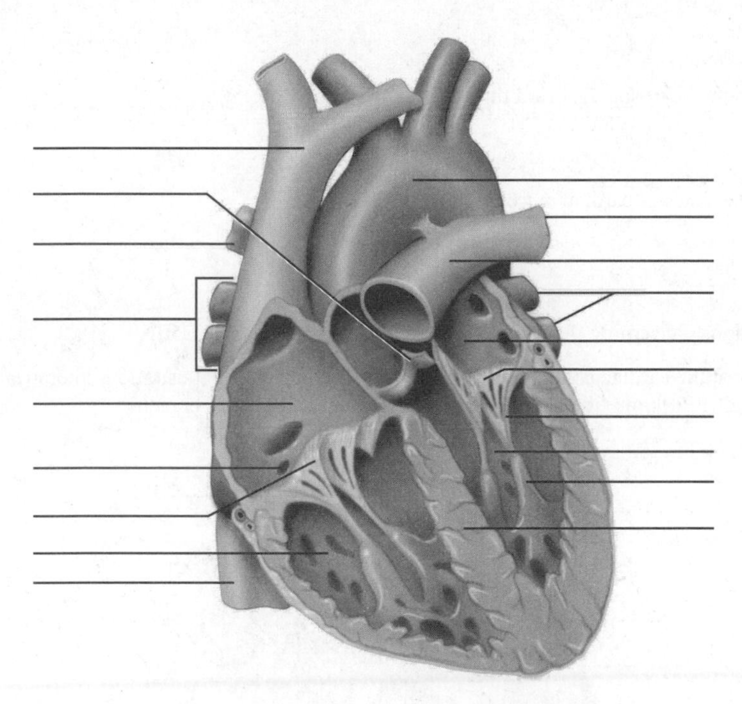

3. What vessels take blood to the right atrium?

4. What vessels take blood to the left atrium?

E. What is mitral valve prolapse? (p. 345)

F. Trace the path of the blood through the heart. Include all valves. (p. 346)

G. The cells of the heart are supplied with blood via the ______________________________. (p. 346)

H. How does the heart maintain a constant supply of oxygenated blood to itself? (p. 347)

I. Compare myocardial infarction and angina pectoris. (p. 347)

13.3 Heart Actions (pp. 347–353)

A. Answer these questions concerning the cardiac cycle. (pp. 347–349)

1. What events make up a cardiac cycle?

2. What produces the heart sounds heard through a stethoscope?

B. Describe the characteristics of cardiac muscle fibers. (p. 349)

C. Answer these questions concerning the cardiac conduction system. (pp. 349–350)

1. Label the parts of the cardiac conduction system in the accompanying illustration: interatrial septum, S-A node, A-V node, A-V bundle, Purkinje fibers, interventricular septum, left bundle branch.

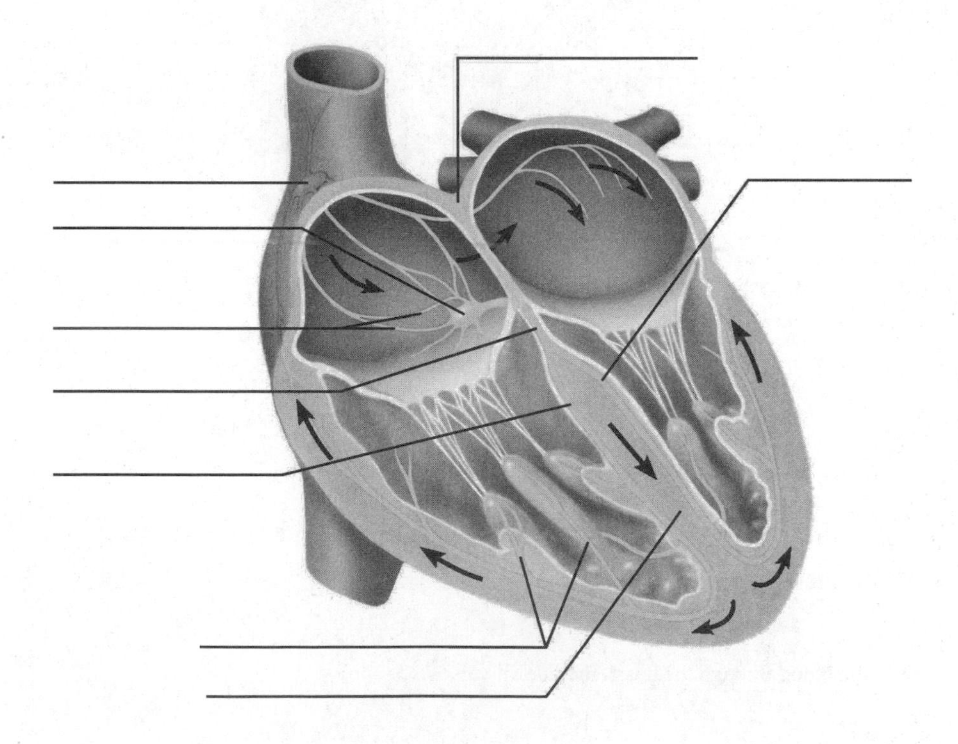

2. Trace an impulse through the cardiac conduction system.

D. Answer these questions concerning the electrocardiogram. (pp. 351–352)

1. What is an electrocardiogram (ECG)?

2. What events in the cardiac cycle are represented by each of the following: P wave, QRS complex, T wave?

E. Answer these questions concerning regulation of the cardiac cycle. (pp. 352–353)

1. Describe how the sympathetic and parasympathetic systems regulate heart function.

2. What is the baroreceptor reflex?

3. How is the heart affected by changes in potassium and calcium levels?

13.4 Blood Vessels (pp. 353–359)

A. Answer these questions concerning the cardiovascular system, arteries, and arterioles. (pp. 353–356)

1. Describe the closed circuit formed by blood vessels.

2. Describe the wall of a typical artery

3. How is the structure of an arteriole different from that of an artery?

4. What controls the diameter of the arterioles and what do changes in the diameter affect?

5. Describe the changes in arteries that occur with atherosclerosis. (p. 355)

B. Answer these questions concerning capillaries. (pp. 356–358)

1. Describe the structure of capillaries.

2. What determines the density of capillaries within tissue?

3. How is the distribution of blood in the various capillary pathways regulated?

4. Describe the following transport mechanisms in the capillary.

   filtration

   osmosis

   diffusion

C. Answer these questions concerning veins and venules. (p. 358)

1. Describe the wall of a vein.

2. In what ways is this different from the wall of an artery?

3. Veins function as blood reservoirs. Why is this important?

13.5 Blood Pressure (pp. 359–363)

A. Answer these questions concerning arterial blood pressure. (p. 359)

1. What is blood pressure?

2. What cardiac events are related to systolic and diastolic arterial pressure?

3. What is the pulse?

B. How do each of the following factors influence arterial blood pressure? (p. 360)

heart action

blood volume

peripheral resistance

viscosity

C. Answer these questions concerning control of blood pressure. (pp. 361–362)

1. How is cardiac output calculated?

2. Discuss the mechanical, neural, and chemical factors that affect cardiac output.

3. How is peripheral resistance regulated?

D. How do factors such as skeletal muscle contraction, breathing, movements, and vasoconstriction of veins influence venous blood flow? (pp. 362–363)

13.6 Paths of Circulation (p. 363)

A. Trace a drop of blood through the pulmonary circuit.

B. Trace a drop of blood through the systemic circuit.

13.7 Arterial System (p. 363–368)

Identify the major vessels of the arterial system. Check with the instructor regarding which vessels to focus on.

13.8 Venous System (p. 369–372)

Identify the major vessels of the venous system. Check with the instructor regarding which vessels to focus on.

Clinical Focus Questions

Your father has just returned from his annual physical and says, “I have nothing to worry about. My heart is fine. I’m just a little overweight, and my blood pressure is a little high.” How would you respond to his comments?

A. What information about high blood pressure is important for your father to understand?

B. What lifestyle changes may be helpful for your father to control his blood pressure?

When you have finished the study activities to your satisfaction, retake the mastery test and compare your results with your initial attempt. If you are not satisfied with your performance, repeat the appropriate study activities.

CHAPTER 14

LYMPHATIC SYSTEM AND IMMUNITY

OVERVIEW

The lymphatic system has two major functions: it helps maintain fluid balance in the tissues of the body, and it has a major role in the defense against infection. This chapter describes the general functions of the lymphatic system (learning outcome 1). In the discussion of fluid balance, it describes the major lymphatic pathways and lymph formation and circulation (learning outcomes 2–4). In describing the defense function, it explains lymph nodes and their functions (learning outcome 5). Innate and adaptive immunity and the functions of the thymus, spleen, lymphocytes, and immunoglobulins are discussed (learning outcomes 6–10). Various types of immune responses—primary and secondary responses, active and passive responses, allergic reactions, tissue rejection reactions and formation, and activation and immune function of lymphocytes—are also explained (learning outcomes 9–13).

Study of the lymphatic system completes the knowledge of how fluid is transported to and away from tissues. Knowledge of the immune mechanisms of the lymphatic system is the basis for understanding how the body defends itself against specific kinds of threats.

LEARNING OUTCOMES

After you have studied this chapter, you should be able to

14.1 Introduction

1. Describe the general functions of the lymphatic system.

14.2 Lymphatic Pathways

2. Identify the locations of the major lymphatic pathways.

14.3 Tissue Fluid and Lymph

3. Describe how tissue fluid and lymph form, and explain the function of lymph.

14.4 Lymph Movement

4. Explain how lymphatic circulation is maintained.

14.5 Lymph Nodes

5. Describe a lymph node and its major functions.

14.6 Thymus and Spleen

6. Discuss the locations and functions of the thymus and spleen.

14.7 Body Defenses Against Infection

7. Distinguish between innate (nonspecific) and adaptive (specific) defenses.

14.8 Innate (Nonspecific) Defenses

8. List seven innate body defense mechanisms, and describe the action of each mechanism.

14.9 Adaptive (Specific) Defenses or Immunity

9. Explain how two major types of lymphocytes are formed and activated and how they function in immune mechanisms.
10. Discuss the actions of the five types of immunoglobulins.
11. Distinguish between primary and secondary immune responses.
12. Distinguish between active and passive immunity.
13. Explain how allergic reactions, tissue rejection reactions, and autoimmunity arise from immune mechanisms.

FOCUS QUESTION

Your entire family has had bronchitis for the past week, but despite close contact with them you have not become ill. How has your lymphatic system helped you to avoid contracting this infection?

MASTERY TEST

Now take the mastery test. Do not guess. As soon as you complete the test, correct it. Note your successes and failures so that you can read the chapter to meet your learning needs.

1. Excess fluid in interstitial spaces is carried away by ____________ vessels.

2. The smallest vessels in the lymphatic system are called ______________________________.
   The largest vessels are called ____________________________________.

3. The largest lymph vessel is the
   a. lumbar trunk.
   b. thoracic duct.
   c. lymphatic duct.
   d. intestinal trunk.

4. Lymph rejoins the blood and becomes part of the plasma in the
   a. lymph nodes.
   b. right and left subclavian veins.
   c. inferior and superior venae cavae.
   d. right atrium.

5. Tissue fluid originates from
   a. the cytoplasm of cells.
   b. lymph fluid.
   c. blood plasma.

6. The function(s) of lymph is (are) to
   a. recapture small protein molecules.
   b. form tissue fluid.
   c. transport foreign particles to lymph nodes.
   d. recapture electrolytes.

7. The mechanisms that move lymph through lymph vessels are similar to those that move blood through (arteries, veins).

8. Obstruction of the flow of lymph leads to ____________________.

9. Lymph nodes are shaped like
   a. almonds.
   b. peas.
   c. beans.
   d. convex discs.

10. Lymph nodes contain dense masses of
    a. epithelial tissue.
    b. cilia.
    c. oocytes.
    d. lymphocytes.

11. The thymus is located in the
    a. posterior neck.
    b. thorax.
    c. upper abdomen.
    d. left pelvis.

12. The thymus produces a substance called ____________________, which seems to stimulate the maturation of ____________lymphocytes.

13. The largest of the lymphatic organs is the ____________________.

14. Which of the following statements about the spleen is (are) true?
    a. The spleen is located in the lower left quadrant of the abdomen.
    b. The spleen functions in the body's defense against infection and as a reservoir for blood.
    c. The structure of the spleen is exactly like that of a lymph node.
    d. Splenic pulp contains large phagocytes on the lining of its venous sinuses.

15. Agents that enter the body and cause disease are called ____________________.

16. The skin is an example of which of the following body defenses against infection?
    a. immunity
    b. inflammation
    c. mechanical barrier
    d. phagocytosis

17. Which of the following characteristics of the stomach enable(s) it to act as a chemical barrier?
a. low pH
b. presence of lysozyme
c. presence of amylase
d. presence of pepsin

18. List the four major symptoms of inflammation.

19. Phagocytes that remain fixed in position within various organs are called
a. neutrophils.
b. monocytes.
c. macrophages.
d. lymphocytes.

20. The resistance to specific foreign agents in which certain cells recognize the foreign substances and act to destroy them is called________________________________.

21. Some undifferentiated lymphocytes migrate to the ____, where they undergo changes and are then called T lymphocytes.

22. Foreign substances to which lymphocytes respond are called ________________________.

23. T cells and B cells seem to be able to recognize specific foreign proteins because
a. of changes in the nucleus of these lymphocytes.
b. the cytoplasm of the T cell and B cell is altered.
c. there are changes in the permeability of the cell membrane of these lymphocytes.
d. of the presence of receptor molecules on T cells and B cells that fit the molecules of antigens.

24. T cells attach themselves to antigen-bearing cells and interact directly in a response called ________________________.

25. B lymphocytes respond to a foreign protein by
a. phagocytosis.
b. interacting directly with pathogens.
c. producing antigens.
d. producing antibodies.

26. The three most abundant types of immunoglobulins are ______________________, ______________________, and ______________________.

27. Antibodies can react to antigens by activating a set of proteins called ____________________ to attack the antigens.

28. T cells require the presence of a(n) ____ before they can become activated.

29. In which of the following ways are primary and secondary immune responses different?
a. Primary responses are more important than secondary responses.
b. Primary responses produce more antibodies than secondary responses.
c. A primary response is a direct response to an antigen; a secondary response is indirect.
d. A primary response is the initial response to an antigen; a secondary response is all subsequent responses to that antigen.

30. A person who receives ready-made antibodies develops artificially acquired ______________________ immunity.

31. An individual with special abilities to carry on abnormal immune reactions against usually harmless substances is said to have
a. an increased potential for cancer.
b. an allergic reaction.
c. a collagen disease.

32. The three types of immune cells targeted by the human immunodeficiency virus are ________________________________, ________________________________, and ________________________________.

33. A common problem following organ transplant is ________________________________.

34. When the immune system fails to distinguish self from nonself, the attack against self is called ________________________.

STUDY ACTIVITIES

Aids to Understanding Words (p. 378)

Define the following word parts. (p. 378)

-gen

humor-

immun-

inflamm-

nod-

patho-

14.1 Introduction (p. 378)

A. Describe the similarities between the lymphatic system and the cardiovascular system.

B. What are the functions of the lymphatic system?

14.2 Lymphatic Pathways (pp. 378–380)

A. What are the components of a lymphatic pathway? (pp. 378–379)

B. Lymph enters the venous system and becomes part of plasma just before the ____. (p. 379)

14.3 Tissue Fluid and Lymph (p. 380)

A. How is tissue fluid formed? Include the composition of tissue fluid and transport mechanisms used.

B. How is lymph formed, and how is lymph formation related to tissue fluid formation?

C. What is the function of lymph?

14.4 Lymph Movement (p. 381)

A. Describe the forces responsible for the circulation of lymph.

B. Trace the flow of lymph from the lymph capillaries to the subclavian veins.

C. What causes edema?

14.5 Lymph Nodes (pp. 381–382)

A. Label these structures in the accompanying illustration (p. 381): afferent vessel, subcapsule, trabecula, germinal center, medulla, sinus, nodule, efferent vessel, hilum, capsule, artery, vein. In addition, indicate the direction of lymph flow through the lymph node.

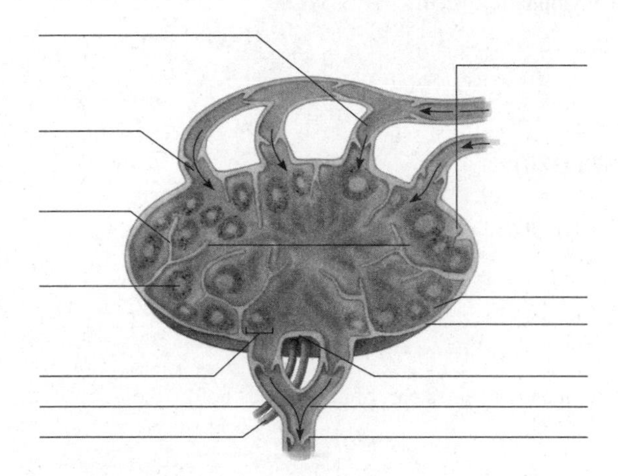

B. Where are lymph nodes located? (p. 382)

C. What is the function of lymph nodes? Include the types of cells found there. (p. 382)

D. Inflammation of the lymph nodes is called ________________________. (p. 382)

14.6 Thymus and Spleen (pp. 382–384)

A. Describe the location and function of the thymus gland. (p. 382)

B. Answer these questions concerning the spleen. (pp. 383–384)

1. Where is the spleen located?

2. Label these structures in the accompanying illustration (p. 384): spleen, capsule, capillary, connective tissue, venous sinus, artery of pulp, red pulp, white pulp, splenic artery, splenic vein.

3. What characteristics of the spleen allow it to filter blood?

4. What characteristics of the spleen allow it to function in the defense against foreign particles?

14.7 Body Defenses Against Infection (p. 384)

A. What kinds of agents cause disease?

B. What two major types of defenses prevent disease?

14.8 Innate (Nonspecific) Defenses (pp. 384–386)

A. What is species resistance? (p. 384)

B. What structures function as mechanical barriers? (p. 385)

C. What chemicals help us resist infection? (p. 385)

D. What is interferon, and how does it work? (p. 385)

E. Answer these questions concerning inflammation. (p. 385)

1. What is inflammation?

2. Explain the reason for the major symptoms of inflammation.

   redness

   swelling

   heat

   pain

3. Explain how inflammation is a defense against infection.

F. What is phagocytosis and what cells are phagocytes? (p. 385)

G. How does fever help protect the body from pathogens? (pp. 385–386)

14.9 Adaptive (Specific) Defenses or Immunity (pp. 386–395)

A. Answer these questions concerning specific immunity and the origin of lymphocytes. (p. 386)

1. What is specific immunity?

2. What are antigens?

3. What is a hapten?

4. Where do lymphocytes originate?

5. Why are some lymphocytes called T lymphocytes and others are called B lymphocytes?

B. Answer these questions concerning T lymphocytes and the cellular immune response. (pp. 386–388)

1. What are antigen-presenting cells? Include the types of cells that can present.

2. Describe how T lymphocytes are activated.

3. What is the cellular immune response?

4. What are T helper cells and how do they assist cell-mediated immunity?

5. What are cytotoxic T cells and how do they function in cell-mediated immunity?

6. What are memory T cells and what is their function?

C. Answer these questions concerning B cells and the humoral immune response. (pp. 390–392)

1. Describe how B lymphocytes are activated.

2. What is the humoral immune response?

3. What are plasma cells? What is an immunoglobulin or antibody?

4. How do antibodies react to antigens?

5. Fill in the following table regarding the classes of immunoglobulins.

Immunoglobulin	**Location**	**Function**
Immunoglobulin G		
Immunoglobulin A		
Immunoglobulin M		
Immunoglobulin D		
Immunoglobulin E		

D. Answer these questions concerning primary and secondary specific immune responses. (pp. 392–393)

1. Define *primary immune response*.

2. Define *secondary immune response*.

E. Define these terms concerning practical classifications of immunity. (p. 393)

naturally acquired active immunity

artificially acquired active immunity

artificially acquired passive immunity

naturally acquired passive immunity

F. Describe the events of an allergic reaction. (pp. 393–394)

1. What is an immediate-reaction allergy?

2. What is a delayed-reaction allergy?

G. How is a tissue rejection reaction an immune response? (p. 394)

H. What is autoimmunity and how does it develop? (pp. 394–395)

Clinical Focus Question

AIDS is a communicable disease that has serious economic, moral, and ethical implications. Discuss how this disease attacks the immune system. Identify your personal perspectives on the economic, moral, and ethical issues related to AIDS.

When you have finished the study activities to your satisfaction, retake the mastery test and compare your results with your initial attempt. If you are not satisfied with your performance, repeat the appropriate study activities.

UNIT 5
ABSORPTION AND EXCRETION
CHAPTER 15

DIGESTION AND NUTRITION

OVERVIEW

This chapter is about the digestive system, which processes food so that nutrients can be absorbed and utilized by cells. It names the organs of the digestive system and describes their locations and functions (learning outcomes 1 and 2). The structure, digestive mechanisms, and movements of the digestive system including defecation and swallowing are explained (learning outcomes 3–6 and 9). The enzymes secreted by the digestive organs and the salivary glands and how these secretions are regulated (learning outcomes 7 and 8) as well as the process of nutrient absorption (learning outcome 10) are discussed. Finally, the nutrients needed by the body, how they are utilized by the body, and the components of an adequate diet are addressed (learning outcomes 11–15).

LEARNING OUTCOMES

After you have studied this chapter, you should be able to

15.1 Introduction

1. Describe the general functions of the digestive system.
2. Name the major organs of the digestive system.

15.2 General Characteristics of the Alimentary Canal

3. Describe the structure of the wall of the alimentary canal.
4. Explain how the contents of the alimentary canal are mixed and moved.

15.3 Mouth

5. Describe the functions of the structures associated with the mouth.
6. Describe how different types of teeth are adapted for different functions, and list the parts of a tooth.

15.4–15.10 Salivary Glands–Large Intestine

7. Identify the function of each enzyme secreted by the digestive organs.
8. Describe how digestive secretions are regulated.
9. Describe the mechanism of swallowing and defecating.
10. Explain how the products of digestion are absorbed.

15.11 Nutrition and Nutrients

11. List the major sources of carbohydrates, lipids, and proteins.
12. Describe how cells utilize dietary carbohydrates, lipids, and proteins.
13. Identify the functions of each fat-soluble and water-soluble vitamin.
14. Identify the functions of each major mineral and trace element.
15. Describe an adequate diet.

FOCUS QUESTION

How does the body make the nutrients in a ham sandwich available for absorption by the cells?

MASTERY TEST

Now take the mastery test. Do not guess. As soon as you complete the test, correct it. Note your successes and failures so that you can read the chapter to meet your learning needs.

1. Digestion is the mechanical and chemical breakdown of foods and the absorption of the resulting nutrients by cells.
   a. True
   b. False

2. The mouth, pharynx, esophagus, stomach, and large and small intestines make up the ________________ of the digestive system.

3. The salivary glands, liver, gallbladder, and pancreas are considered ________________ ________________.

4. The layer of the wall of the alimentary canal that is formed of surface epithelium and protects underlying tissue while carrying on absorption and secretion is the
   a. mucosa.
   b. submucosa.
   c. muscular layer.
   d. serosa.

5. The layer of the alimentary tube that keeps the outer surface of the alimentary tube moist and slippery is the
   a. mucosa.
   b. submucosa.
   c. muscular layer.
   d. serosa.

6. The two basic types of movement of the alimentary canal are________________ movements and ________________ movements.

7. Does statement a explain statement b? (Yes, No)
   a. Peristalsis is stimulated by stretching the alimentary tube.
   b. Peristalsis acts to move food along the alimentary tube.

8. The tongue is anchored to the floor of the mouth by a fold of membrane called the ________________.

9. The material that covers the crown of the teeth is
   a. cementum.
   b. dentin.
   c. enamel.
   d. plaque.

10. Teeth that are chisel-shaped and whose function is to bite off pieces of food are
    a. incisors.
    b. cuspids.
    c. bicuspids.
    d. molars.

11. What parts of a tooth are destroyed by dental caries?
    a. enamel
    b. dentin
    c. pulp
    d. cementum

12. Which of the following is *not* a function of saliva?
    a. cleanses mouth and teeth
    b. dissolves chemicals necessary to tasting food
    c. helps in formation of food bolus
    d. begins digestion of protein

13. Stimulation of salivary glands by parasympathetic nerves will (increase, decrease) production of saliva.

14. The salivary glands that secrete amylase are the
    a. submaxillary glands.
    b. parotid glands.
    c. sublingual glands.

15. During swallowing, muscles draw the soft palate and uvula upward to
    a. move food into the esophagus.
    b. enlarge the area to accommodate a bolus of food.
    c. separate the oral and nasal cavities.
    d. move the uvula from the path of the food bolus.

16. When food enters the esophagus, it is transported to the stomach by a movement called ________________.

17. The area of the stomach that acts as a temporary storage area is the
a. cardiac region.
b. fundic region.
c. body region.
d. pyloric region.

18. The chief cells of the gastric glands secrete
a. mucus.
b. hydrochloric acid.
c. digestive enzymes.
d. potassium chloride.

19. The digestive enzyme pepsin secreted by gastric glands begins the digestion of
a. carbohydrates.
b. protein.
c. fats.

20. The intrinsic factor secreted by the stomach aids in the absorption of ______________________________ from the small intestine.

21. The release of gastrin is stimulated by the
a. parasympathetic nervous system.
b. presence of alkaline substances.
c. sight and smell of food.
d. presence of such substances as protein, caffeine, and alcohol.

22. Gastric ulcers are considered to be
a. a disease of stress.
b. an infectious disease.
c. an endocrine disorder.
d. a product of overactive parietal cells.

23. The presence of food in the small intestine (inhibits, increases) gastric secretion.

24. The semifluid paste formed in the stomach by mixing food and gastric secretions is called______________________________.

25. The foods that stay in the stomach the longest are high in
a. fats.
b. protein.
c. carbohydrates.

26. Pancreatic enzymes travel along the pancreatic duct and empty into the
a. duodenum.
b. jejunum.
c. ileum.

27. Which of the following enzymes is (are) present in secretions of the mouth, stomach, and pancreas?
a. amylase
b. lipase
c. trypsin
d. lactase

28. Which of the following is (are) secreted by the pancreas in an inactive form and is (are) activated by a duodenal enzyme?
a. nuclease
b. trypsin
c. chymotrypsin
d. carboxypeptidase

29. The secretions of the pancreas are (acid, alkaline).

30. The liver is located in the____ quadrant of the abdomen.

31. The liver's most vital functions are related to metabolism of
a. carbohydrates.
b. fats.
c. cholesterol.
d. protein.

32. Type A hepatitis is transmitted by
a. ingestion of food contaminated by feces.
b. transfusion with blood contaminated by the hepatitis virus.
c. use of improperly cleaned needles.
d. sexual activity.

33. Which of the following substances is (are) not stored in the liver?
a. vitamins A and D
b. protein
c. iron
d. water-soluble vitamins

34. Phagocytic cells found in the inner linings of the sinusoids of the liver are ______________________________ cells.

35. The only substances in bile that have a digestive function are ______________________.

36. The function(s) of the gallbladder is (are) to
a. store bile.
b. secrete bile.
c. activate bile.
d. concentrate bile.

37. The gallbladder is stimulated to release bile by the hormone ______________________.

38. Which of the following is (are) the function(s) of bile?
a. emulsification of fat globules
b. absorption of fats
c. increase of solubility of amino acids
d. absorption of fat-soluble vitamins

39. List the parts of the small intestine: ______________________, ______________________, and ______________________.

40. The velvety appearance of the lining of the small intestine is due to the presence of
a. cilia.
b. villi.
c. mucus secreted by the small intestine.
d. capillaries.

41. The small intestine absorbs (most, few) of the products of digestion.

42. Digestive enzymes and mucus (are, are not) secreted by the small intestine.

43. Peristaltic rush in the small intestine results in ______________________.

44. The small intestine joins the large intestine at the ______________________.

45. The only significant secretion of the large intestine is
a. potassium.
b. mucus.
c. chyme.
d. water.

46. The only nutrients normally absorbed in the large intestine are ______________________ and ______________________.

47. The defecation reflex can be initiated by
a. holding a deep breath.
b. seeing and smelling food.
c. contracting the abdominal wall muscles.
d. sensing fullness in the abdomen.

48. The most abundant substance in feces is ______________________.

49. Nutrients, such as amino acids and fatty acids, that are necessary for health but cannot be synthesized in adequate amounts by the body are called ______________________.

50. Carbohydrates are ingested in such foods as
a. meat and seafood.
b. bread and pasta.
c. butter and margarine.
d. bacon.

51. A carbohydrate that cannot be broken down by human digestive enzymes and facilitates muscle activity in the alimentary tube is ______________________.

52. Glucose can be stored as glycogen in the
a. blood plasma.
b. muscles.
c. connective tissue.
d. liver.

53. The organ most dependent upon an uninterrupted supply of glucose is the
a. heart muscle.
b. liver.
c. adrenal gland.
d. brain.

54. An essential fatty acid that cannot be synthesized by the body is ______________________.

55. A lipid that furnishes molecular components for the synthesis of sex hormones and some adrenal hormones is ______________________.

56. Proteins function as
a. enzymes that regulate metabolic reactions.
b. promoters of calcium absorption.
c. energy supplies.
d. structural materials in cells.

57. Proteins are absorbed and transported to cells as ______________________.

58. A protein that contains adequate amounts of the essential amino acids is called a ______________________ protein.

59. The least stable vitamin is vitamin ________________.

60. The vitamin that promotes absorption of calcium and phosphorus is vitamin ________________.

61. The most abundant minerals in the body are ________________ and ________________.

62. Iron is associated with the body's ability to transport ________________.

STUDY ACTIVITIES

Aids to Understanding Words

Define the following word parts. (p. 401)

aliment-

chym-

decidu-

gastr-

hepat-

lingu-

nutri-

peri-

pyl-

vill-

15.1 Introduction (p. 401)

A. Define *digestion.*

B. Label this illustration of the digestive system. (p. 402)

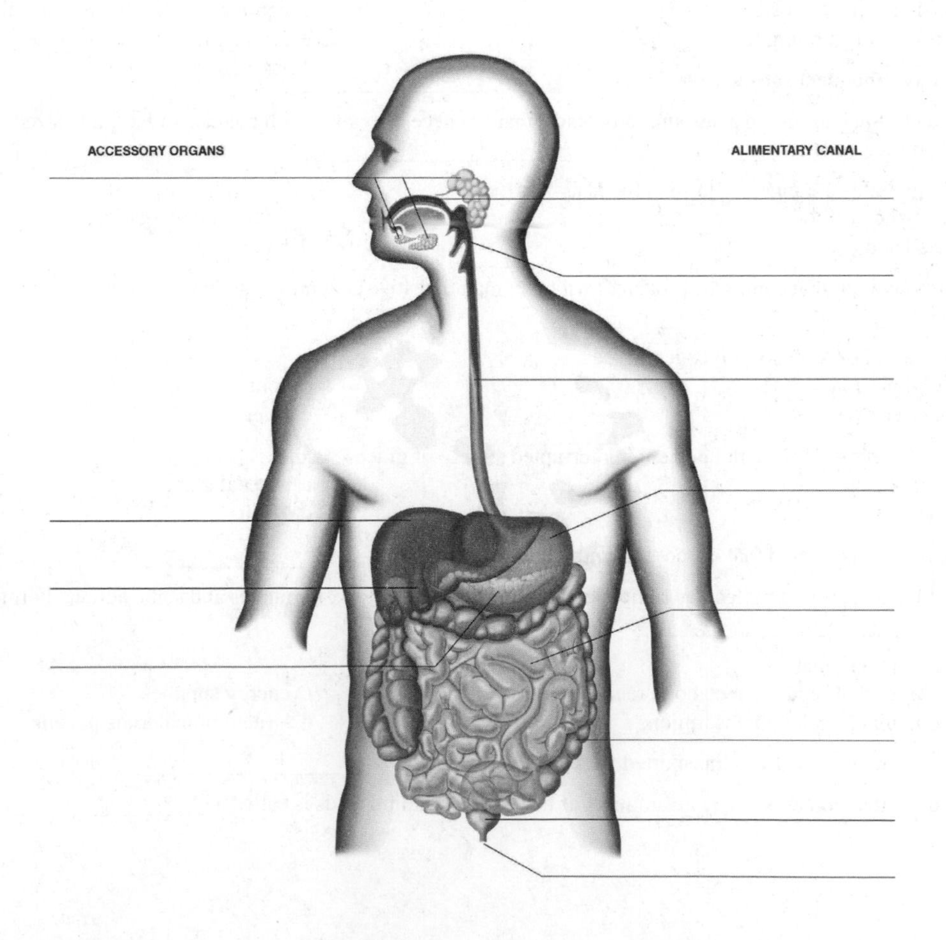

15.2 General Characteristics of the Alimentary Canal (pp. 401–403)

A. Fill in the following table regarding the tissue layers of the alimentary canal. (p. 401)

Layer	Composition	Function
Mucous membrane		
Submucosa		
Muscular		
Serosa		

B. The two types of movement of the alimentary tube are ______________________ and ______________________. What is peristalsis? (p. 403)

15.3 Mouth (pp. 403–408)

A. What are the structures and functions of the mouth? (p. 403)

B. What are the functions of the tongue? (pp. 403–404)

C. Answer these questions concerning the palate. (pp. 404–405)

1. What are the parts of the palate?

2. What is the function of the palatine tonsils?

D. In the accompanying illustration (p. 408), label and assign a function to the following structures: the crown, root, enamel, dentin, gingiva, pulp cavity, cementum, alveolar bone, root canal, periodontal ligament.

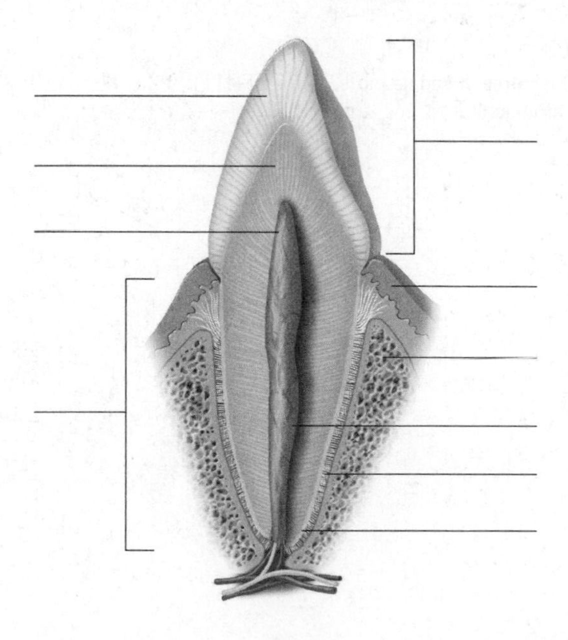

E. How can individuals best care for their teeth? (p. 407)

15.4 Salivary Glands (p. 408)

A. Answer these questions concerning salivary glands and their secretions. (p. 408)

1. What is the function of salivary glands?

2. What is the composition of saliva?

B. Fill in the following table regarding the salivary glands. (p. 408)

Glands	Location of the glands	Secretion
Parotid		
Submandibular		
Sublingual		

15.5 Pharynx and Esophagus (pp. 408–410)

A. Describe the nasopharynx, oropharynx, and laryngopharynx. (p. 409)

B. List the actions of the swallowing reflex. Include the function of the soft palate in this process. (p. 409)

C. Describe the structure and functions of the esophagus. (p. 410)

15.6 Stomach (pp. 410–413)

A. Answer this question concerning the stomach and its parts. (pp. 410–411)

1. What are the functions of the stomach?

2. Label the following structures (p. 410): fundic region, cardiac region, body, pyloric region, pyloric canal, duodenum, pyloric sphincter, rugae, esophagus.

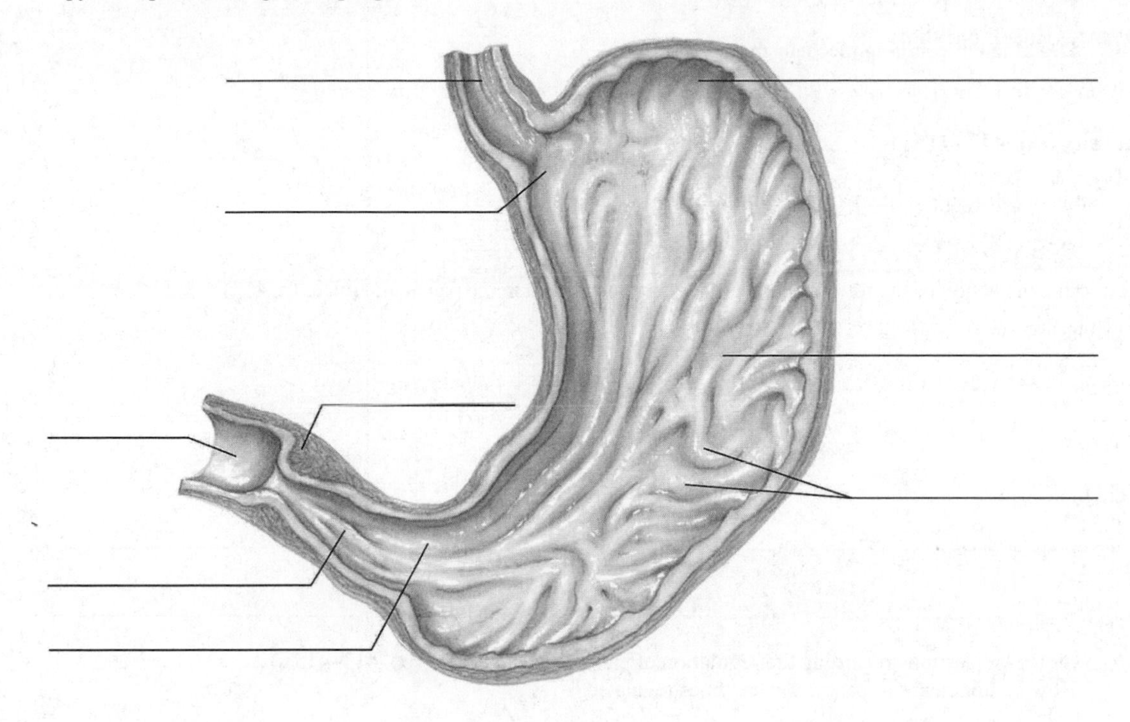

B. Fill in the following table regarding secretions of the gastric glands. (pp. 411–412)

Cell type	Secretions	Function and action
Mucous cell		
Chief cell		
Parietal cell		

C. Answer these questions regarding regulation of gastric secretions. (p. 412)

1. What are the neural controls of gastric secretion?

2. What are the hormonal controls of gastric secretion?

D. What substances are absorbed from the stomach? (p. 412)

E. Answer the following questions regarding the mixing and emptying actions. (p. 413)

1. What is chyme, and how is it produced?

2. What factors affect the rate at which the stomach empties?

3. Describe the vomiting mechanism.

15.7 Pancreas (pp. 413–415)

A. What is the structure of the pancreas and where is it located? (p. 413)

B. Describe the action of the following pancreatic enzymes found in the pancreatic juice. (p. 413)

amylase

lipase

trypsin

chymotrypsin

carboxypeptidase

nuclease

C. Answer these questions regarding the regulation of pancreatic secretions. (pp. 413–415)

1. How is the release of pancreatic enzymes regulated?

2. What substance makes the pancreatic juice alkaline?

3. How does secretin affect pancreatic juice?

4. What is the function of cholecystokinin?

15.8 Liver (pp. 415–419)

A. Answer these questions regarding the liver's structure and function. (pp. 415–417)

1. Describe the location of the liver.

2. The liver is divided into _______________ lobes.

3. The functional units of the liver are the ______________.

4. Describe the structure of a hepatic lobule.

5. The hepatic portal vein takes blood to the liver from the ______________________________________.

6. The large macrophages in the inner linings of the hepatic sinusoids are ____________________.

7. Secretions from the hepatic cells are collected in _________________________, which converge to form _________________________.

8. Describe the digestive function of the liver.

B. Fill in the following table regarding types of hepatitis. (p. 417)

Type of hepatitis	Mode of transmission	Patient characteristics
Hepatitis A		
Hepatitis B		
Hepatitis C		
Hepatitis D		
Hepatitis E		
Hepatitis F		
Hepatitis G		

C. Answer these questions concerning bile. (p. 418)

1. Describe the composition of bile.

2. Which of the substances in bile is active in the digestive process?

3. What is the source of bile pigments?

D. Answer these questions concerning the gallbladder. (p. 418)

1. Describe the structure and location of the gallbladder.

2. How is bile stored?

E. Describe how bile is released. (p. 418)

F. Answer these questions concerning the digestive functions of bile salts. (pp. 418–419)

1. The digestive function of bile is the ______________________ of fats.

2. Bile salts aid the absorption of ______________________, ______________________, and ______________________.

15.9 Small Intestine (pp. 420–424)

A. Answer these questions regarding the parts of the small intestines. (p. 420)

1. Name and locate the three parts of the small intestine.

2. The double layer of peritoneal membrane that suspends the jejunum and the ileum from the posterior wall of the abdomen is the ______________________________.

3. A membrane that can contain and localize infections in the alimentary canal is the ____________________ ______________.

B. Label these structures in the accompanying illustration and explain the function of each (p. 421): lacteal, capillary network, intestinal gland, goblet cells, arteriole, venule, lymph vessel, villus, simple columnar epithelium.

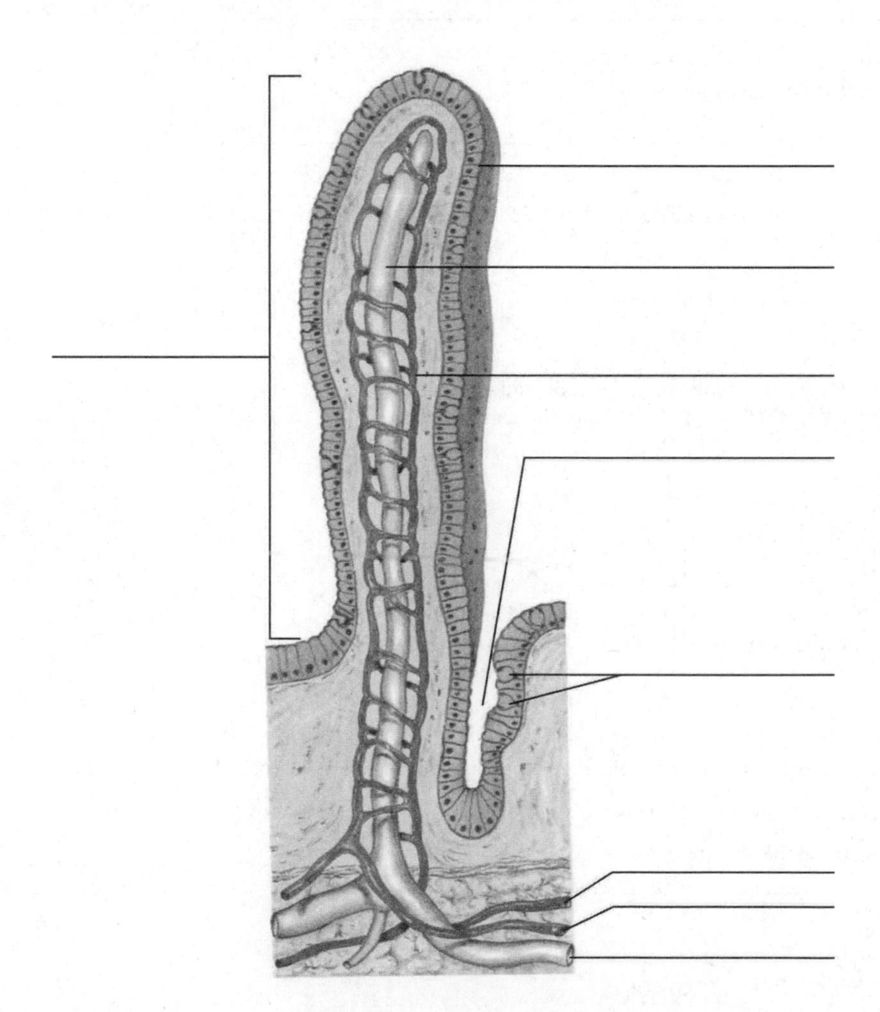

C. List the secretions of the small intestine, and describe their functions. How are these secretions regulated? (pp. 421–422)

D. Answer the following questions regarding absorption in the small intestine. (pp. 422–424)

1. The structures that make absorption in the small intestine so efficient are the ____________.
2. Carbohydrates are absorbed in the small intestine as ____________.
3. Protein is absorbed in the small intestine as ____________ ____________.
4. Fat molecules are encased in protein in the small intestine to form ______________.
5. Fill in the following table regarding the absorption of nutrients.

Nutrient	Absorption mechanism	Means of transport
Monosaccharides		
Amino acids		
Fatty acids & glycerol		
Electrolytes		
Water		

15.10 Large Intestine (pp. 424–428)

A. Answer these questions regarding the large intestine and its parts. (pp. 424–426)

1. Label the parts of the large intestine in the accompanying illustration (p. 425): vermiform appendix, cecum, ascending colon, transverse colon, descending colon, ileocecal sphincter, serous layer, muscular layer, mucous layer, sigmoid colon, rectum, anal canal, tenia coli, haustra, orifice of appendix.

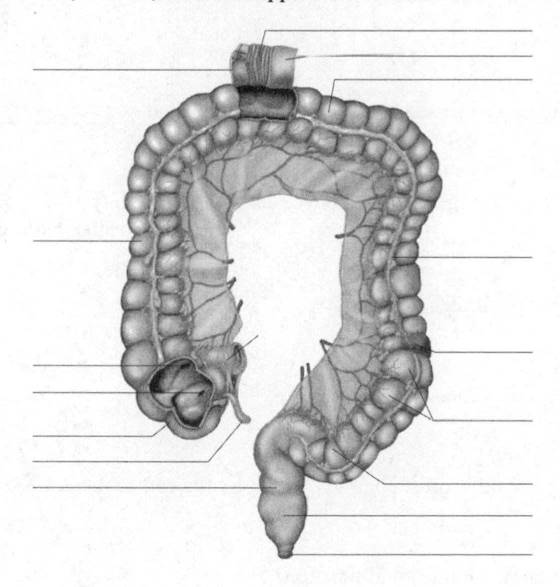

2. How is the structure of the wall of the large intestine different from the structure of the wall of the small intestine?

3. What are hemorrhoids?

B. Answer these questions concerning the functions of the large intestine. (pp. 426–427)

1. What is the function of mucus in the large intestine?

2. What substances are absorbed in the large intestine?

3. What is the role of intestinal bacteria?

C. Answer these questions concerning the movements of the large intestine. (p. 427)

1. Describe the movements of the large intestine.

2. List the events of the defecation reflex.

D. Describe the composition of feces. (p. 427)

15.11 Nutrition and Nutrients (pp. 428–436)

A. Answer the following general questions regarding nutrients. (p. 428)

1. Define *nutrition.*

2. Nutrients that cannot be synthesized by human cells are called ____________________ nutrients.

B. Answer the following questions regarding carbohydrates and their use. (pp. 428–429)

1. Carbohydrates are ______________________________ compounds that are used primarily to supply ______________________________.

2. In what forms are carbohydrates ingested?

3. In what forms are carbohydrates absorbed?

4. What form of carbohydrate is most commonly used by the cell as fuel?

5. How does cellulose facilitate intestinal function?

6. Identify the areas in which glucose is stored for rapid mobilization.

7. What is the function of carbohydrates?

8. What cells are particularly dependent upon a continuous supply of glucose?

9. What is the estimated daily requirement for carbohydrate?

C. Answer the following questions regarding lipids and their use. (pp. 429–430)

1. In what forms are lipids usually ingested?

2. What are the end products of triglyceride digestion?

3. Describe the role of the liver in the utilization of fats.

4. An essential fatty acid that cannot be synthesized by the body is ________________________________ ________________________________.

5. Describe the role of adipose tissue in the utilization of lipids.

6. What is the function of lipids?

7. What is the estimated daily requirement for lipids?

D. Answer the following questions regarding proteins and their use. (pp. 430–432)

1. What are the functions of protein?

2. What is an essential amino acid?

3. What is the difference between a complete and an incomplete protein?

4. Why must various sources of vegetable protein be combined in a meal?

5. What is the estimated daily requirement for protein?

E. Answer the following questions regarding vitamins. (pp. 432–433)

1. What is a vitamin?

2. What are the general characteristics of fat-soluble vitamins?

3. What are the general characteristics of water-soluble vitamins?

4. Fill in the following table regarding important vitamins.

Vitamin	Characteristics	Functions	Sources
Vitamin A			
Vitamin D			
Vitamin E			
Vitamin K			
Ascorbic acid (vitamin C)			
Vitamin B complex			
Thiamin (B_1)			
Riboflavin (B_2)			
Niacin (nicotinic acid)			
Pyridoxine (B_6)			
Pantothenic acid			
Cyanocobalamin (B_{12})			
Folacin (folic acid)			
Biotin			

F. Answer the following questions regarding minerals. (pp. 433–435)

1. Describe the general characteristics of minerals.

2. Fill in the following table regarding the major minerals.

Mineral	**Distribution**	**Regulatory mechanism**	**Functions**	**Sources**
Calcium				
Phosphorus				
Potassium				
Sulfur				
Sodium				
Chlorine				
Magnesium				

3. What is a trace element?

4. List the nine trace elements, and identify the function of each.

G. Answer the following general questions regarding dietary intake. (pp. 434–436)

1. What is an adequate diet?

2. Define *malnutrition, undernutrition*, and *overnutrition.*

Clinical Focus Question

A close friend is planning to be married in three months and tells you she must lose ten pounds before the wedding. She states that she is under a great deal of stress as the wedding approaches. How would you advise her? Be sure to consider the role of exercise and the effects of stress on nutritional needs.

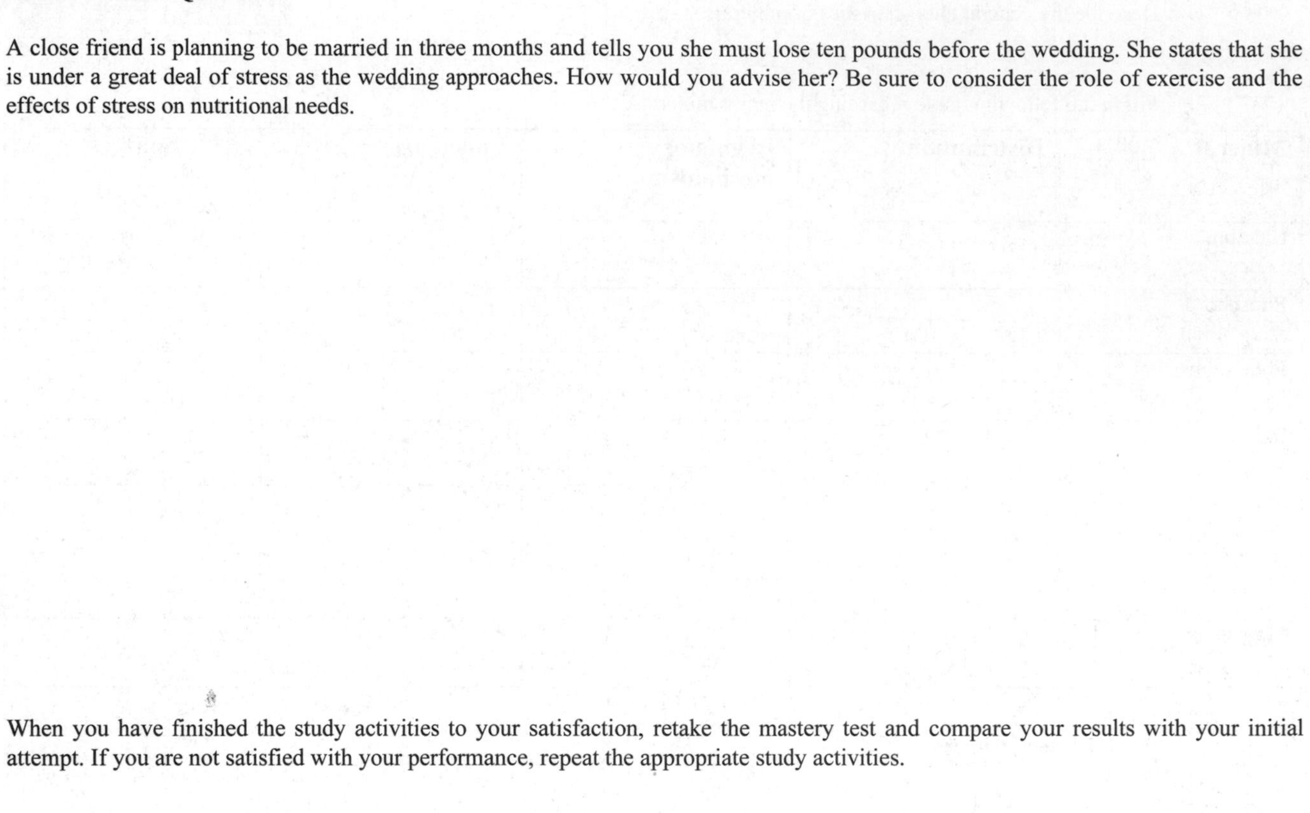

When you have finished the study activities to your satisfaction, retake the mastery test and compare your results with your initial attempt. If you are not satisfied with your performance, repeat the appropriate study activities.

CHAPTER 16

RESPIRATORY SYSTEM

OVERVIEW

The respiratory system permits the exchange of oxygen, which is needed for cellular metabolism, and carbon dioxide, which is produced by cellular metabolism. This chapter describes the location and function of the organs of the respiratory system and discusses how they contribute to the overall function of the system (learning outcomes 1–3). Respiratory air movements, respiratory volumes and capacities, and how normal breathing is controlled are discussed (learning outcomes 4–7). Gas exchange and transport are explained (learning outcomes 8–10).

Air must be taken into the lungs so that oxygen and carbon dioxide can be exchanged. An understanding of these events and how they are controlled is basic to understanding how cells produce energy for life processes.

LEARNING OUTCOMES

After you have studied this chapter, you should be able to

16.1 Introduction

1. Identify the general functions of the respiratory system.

16.2 Organs of the Respiratory System

2. Locate the organs of the respiratory system.
3. Describe the functions of each organ of the respiratory system.

16.3 Breathing Mechanisms

4. Explain the mechanisms of inspiration and expiration.
5. Define each of the respiratory volumes and capacities.

16.4 Control of Breathing

6. Locate the respiratory areas in the brainstem and explain how they control breathing.
7. Discuss how various factors affect the respiratory areas.

16.5 Alveolar Gas Exchanges

8. Describe the structure and function of the respiratory membrane.
9. Explain how air and blood exchange gases.

16.6 Gas Transport

10. List the ways blood transports oxygen and carbon dioxide.

FOCUS QUESTION

You have climbed three flights of steps to get to your classroom. You then sit quietly in a lecture hall, taking notes. How does the respiratory system help maintain your body equilibrium with such different levels of physical activity?

MASTERY TEST

Now take the mastery test. Do not guess. As soon as you complete the test, correct it. Note your successes and failures so that you can read the chapter to meet your learning needs.

1. The process of exchanging gases between the atmosphere and body cells is called ______________________________.

2. Match the functions in the first column with the appropriate part of the nose in the second column.

____a. warm(s) incoming air
____b. trap(s) particulate matter in the air
____c. prevent(s) infection
____d. moisten(s) air
____e. move(s) nasal secretions to pharynx

1. mucous membrane
2. mucus
3. cilia

3. The pharynx is the cavity behind the mouth, extending from the ____________________________ to the ____________________________.

4. The portion(s) of the larynx that prevent(s) foreign objects from entering the trachea is (are) the
a. arytenoid cartilages.
b. glottis.
c. epiglottis.
d. hyoid bone.

5. The trachea is maintained in an open position by
a. cartilaginous rings.
b. the amount of collagen in the wall.
c. the tone of smooth muscle in the wall of the trachea.
d. the continuous flow of air through the trachea.

6. The right and left bronchi arise from the trachea at the
a. suprasternal notch.
b. manubrium of the sternum.
c. fifth thoracic vertebra.
d. eighth intercostal space.

7. The smallest branches of the bronchial tree are the ____.

8. The serous membrane covering the lungs is the ____.

9. The serous membrane covering the inner wall of the thoracic cavity is the ______________________________ ______________________.

10. The right lung is (larger, smaller) than the left lung.

11. The pressure in the thoracic cavity during inspiration is
a. greater than atmospheric pressure.
b. less than atmospheric pressure.
c. the same as atmospheric pressure.

12. Inspiration occurs after the diaphragm ______________________, thus (increasing, decreasing) the size of the thorax and (increasing, decreasing) the pressure within the thorax.

13. The other muscles that normally act to change the size of the thorax are the
a. sternocleidomastoids.
b. pectorals.
c. intercostals.
d. latissimus dorsi.

14. Expansion of the lungs during inspiration is assisted by the surface tension of fluid in the______________________cavity.

15. The surface tension of fluid in the alveoli is decreased by the secretion of ________________________, which prevents collapse of the alveoli.

16. The disease that results in decreased surface area of the respiratory membrane and loss of elasticity in the alveolar walls is ________________________.

17. The force responsible for expiration comes mainly from
a. contraction of intercostal muscles.
b. change in the surface tension within alveoli.
c. elastic recoil of tissues in the lung and thoracic wall.
d. contraction of abdominal muscle to push the diaphragm upward.

18. The amount of air that enters and leaves the lungs during a normal, quiet respiration is known as
a. vital capacity.
b. inspiratory reserve volume.
c. total lung capacity.
d. tidal volume.

19. Normal breathing is controlled by the respiratory areas located in the______________________and the __________________________.

20. The respiratory group that is most important in stimulating the muscles of the diaphragm to contract during inspiration is the
a. pons.
b. pneumotaxic center.
c. dorsal respiratory group.
d. brainstem.

21. The part of the respiratory control system thought to contribute to the basic rhythm of breathing is the ____________________________ group.

22. The inflation reflexes are activated by
a. stretch receptors in the bronchioles and alveoli.
b. an increase in hydrogen ions.
c. a decrease in oxygen saturation.
d. a sudden fall in blood pressure.

23. The strongest stimulus to increase respiratory rate and depth is to increase the blood concentration of ____________________________.

24. The respiratory membrane consists of a single layer of epithelial cells and basement membrane between a(n) ____________________________ and a(n) ____________________________.

25. The rate at which a gas diffuses from one area to another is determined by differences in__________________in the two areas.

26. The pressure of each gas within a mixture is known as its ____________________________.

27. Oxygen is transported to cells by combining with ____________________________.

28. The largest amount of carbon dioxide is transported
a. dissolved in blood.
b. combined with hemoglobin.
c. as bicarbonate.
d. as carbonic anhydrase.

STUDY ACTIVITIES

Aids to Understanding Words

Define the following word parts. (p. 443)

alveol-

bronch-

cric-

epi-

hem-

16.1 Introduction (p. 443)

Define the following terms related to respiration.

1. *Ventilation*

2. *Gas exchange*

3. *Gas transport*

4. *Cellular respiration*

16.2 Organs of the Respiratory System (pp. 443–450)

A. In the accompanying illustration (p. 443), label these structures: nasal cavity, nostril, pharynx, larynx, trachea, bronchus, right lung, left lung, epiglottis, hard palate, soft palate, oral cavity, esophagus, frontal sinus

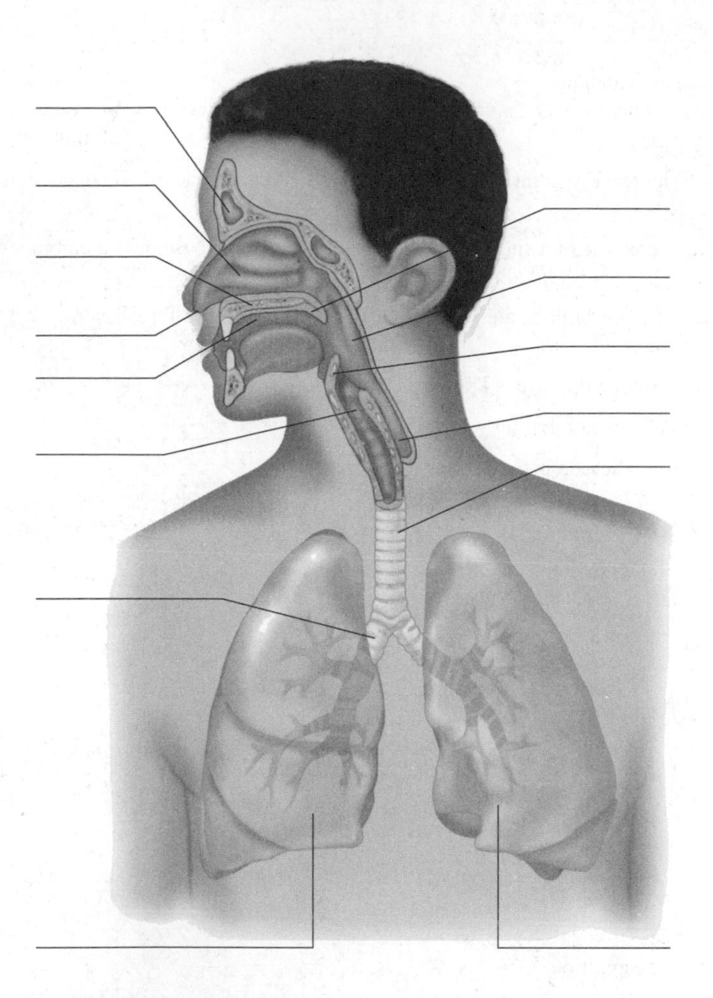

B. Match the functions in the first column with the appropriate terms in the second column. (pp. 443–444)

____1.	entrap(s) dust	a. mucous membrane
____2.	lighten(s) skull and provide(s) vocal resonance	b. mucus
____3.	warm(s) and humidify(ies) air entering the nose	c. sinuses
____4.	provide(s) movement to mucous layer	d. cilia

C. Describe the location and functions of the paranasal sinuses. (p. 444)

D. What is the location and function of the pharynx? (p. 445)

E. Answer these questions concerning the larynx. (pp. 445–446)

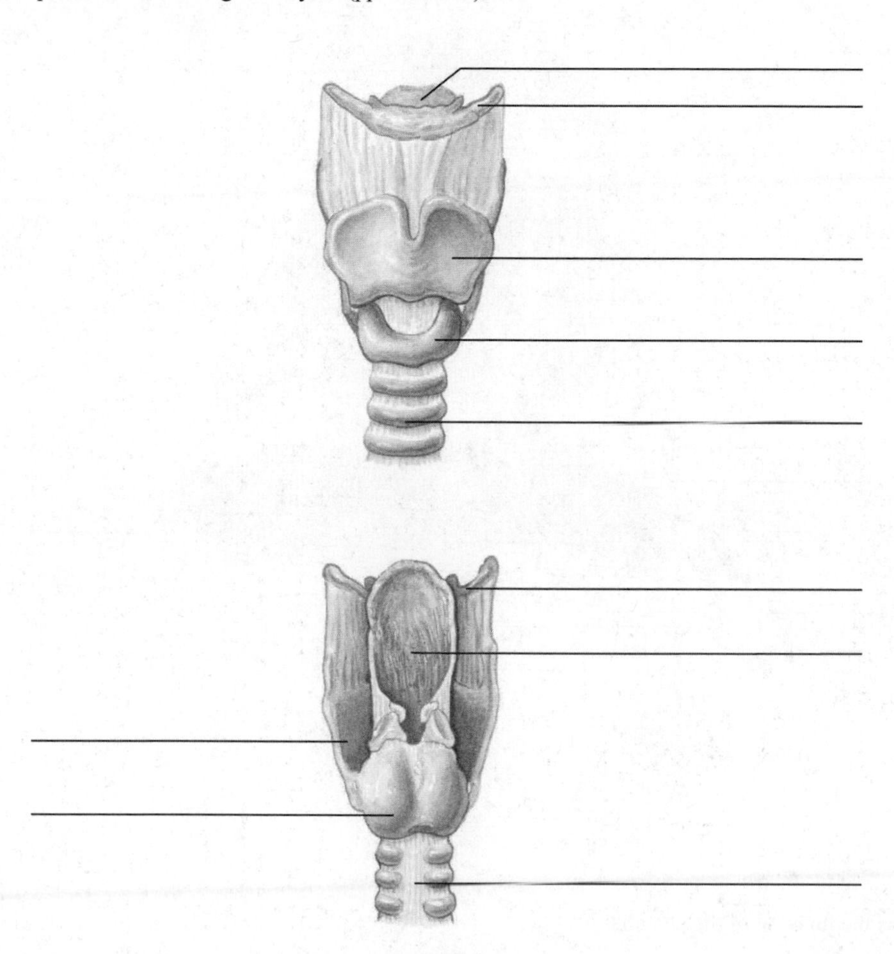

1. Identify these structures in the accompanying illustration (p. 445): hyoid bone, epiglottic cartilage, trachea, thyroid cartilage, cricoid cartilage.
2. The structure of the larynx that helps close the glottis during swallowing is the ______________________________.
3. The structures of the larynx that produce sound are the ______________________________.

F. Describe the structure and function of the inner wall of the trachea. (p. 446)

G. Answer these questions concerning the bronchial tree. (pp. 446–449)

1. In the accompanying illustration (p. 447), label the larynx, trachea, right primary bronchus, secondary bronchus, alveolus, alveolar duct, bronchiole, right superior lobe of lung, right middle lobe of lung, right inferior lobe of lung, left superior lobe of the lobe of lung, left inferior lobe of lung, terminal bronchiole, tertiary branches.

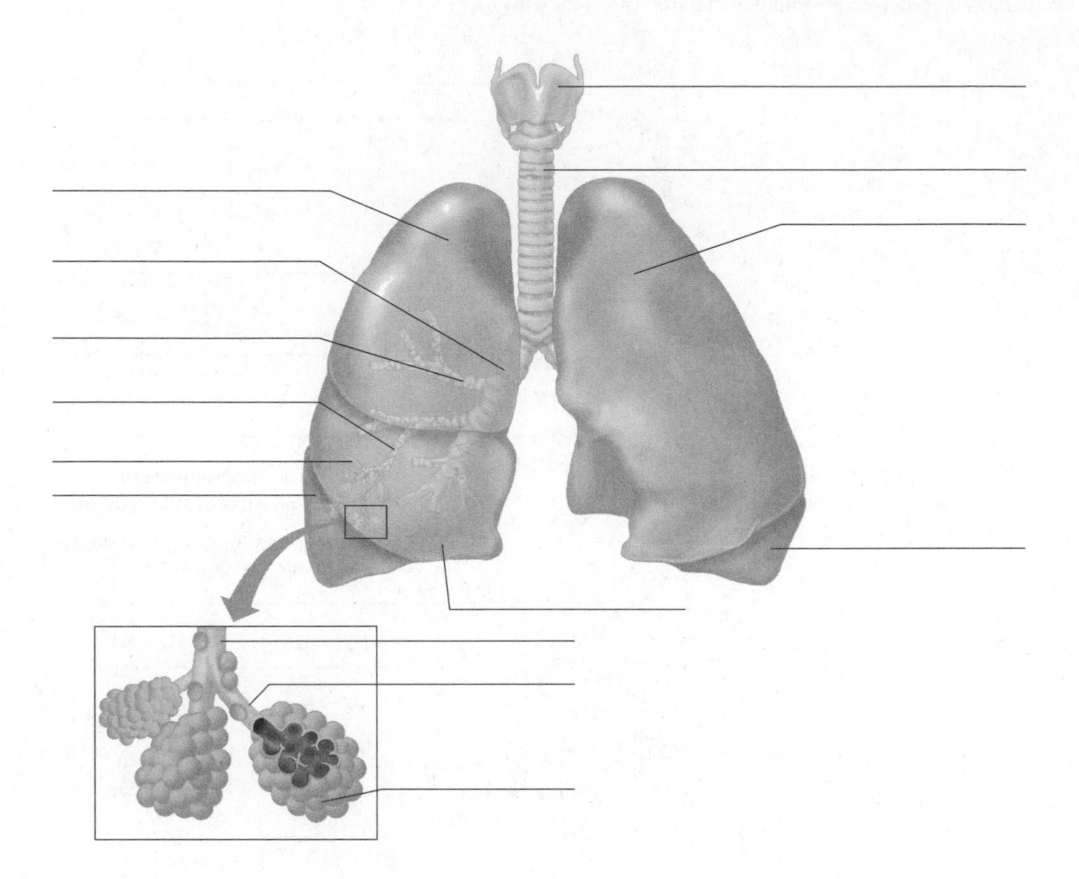

2. What is the function of the alveoli?

H. Answer these questions concerning the lungs. (p. 450)

1. The right lung has ____________________________ lobes, and the left lung has_________________lobes.

2. How is the pleural cavity formed?

16.3 Breathing Mechanisms (pp. 450–456)

A. Answer these questions regarding inspiration. (pp. 450–452)

1. List the events of inspiration.

2. Describe the role of surface tension in the pleural cavity and in the alveoli.

3. How does surfactant support the function of the alveoli?

B. List the events of expiration. (p. 453)

C. Match the terms in the first column with their correct definition in the second column. (pp. 455–456)

____1. inspiratory reserve volume
____2. expiratory reserve volume
____3. residual volume
____4. vital capacity
____5. total lung capacity
____6. tidal volume
____7. inspiratory capacity
____8. functional residual capacity

a. volume of air that remains in the lungs following exhalation of tidal volume
b. volume moved into or out of the lungs during quiet respiration
c. volume that can be inhaled during forced breathing in addition to tidal volume
d. volume that can be exhaled in addition to tidal volume
e. volume that remains in the lungs at all times
f. maximum air that can be exhaled after taking the deepest possible breath
g. total volume of air the lungs can hold
h. maximum volume of air that can be inhaled following exhalation of tidal volume

D. What is anatomical dead space and where is it found? (p. 456)

16.4 Control of Breathing (pp. 456–458)

A. List the neural, muscular, skeletal, and pulmonary structures involved in the control of breathing. (p. 456)

B. Describe the function of the respiratory areas in maintaining normal breathing. Include the medullary respiratory center, dorsal respiratory group, ventral respiratory group, and pontine respiratory group. (p. 456)

C. List the factors that affect breathing and describe how they alter breathing. (pp. 456–458)

16.5 Alveolar Gas Exchanges (pp. 459–460)

A. Identify the layers of cells that constitute the respiratory membrane. (p. 459)

B. Answer these questions concerning diffusion through the respiratory membrane. (pp. 459–460)

1. What determines the direction and rate at which gases diffuse from one area to another?

2. Define *partial pressure of gas*.

3. Use partial pressure of gas to explain the exchange of oxygen and carbon dioxide in the alveoli.

16.6 Gas Transport (pp. 460–462)

A. Answer these questions concerning oxygen transport. (p. 460)

1. Describe how oxygen is transported to cells.

2. Why is oxygen released to the cell?

B. Answer these questions concerning carbon dioxide transport. (pp. 460–462)

1. How is carbon dioxide transported away from the cell?

2. Why is carbon dioxide removed from blood in the lungs?

Clinical Focus Questions

A. How do the anatomical changes of emphysema affect the physiology of the respiratory system? Be sure to consider changes in the mechanism of breathing, gas exchange, and the control of breathing.

B. What differences would you expect when auscultating a normal lung and an emphysemic lung?

When you have finished the study activities to your satisfaction, retake the mastery test and compare your results with your initial attempt. If you are not satisfied with your performance, repeat the appropriate study activities.

CHAPTER 17

URINARY SYSTEM

OVERVIEW

The urinary system plays a vital role in maintenance of the internal environment by excreting nitrogenous waste products and by selectively excreting or retaining water and electrolytes. This chapter identifies, locates, and describes the functions of the organs of the urinary system (learning outcomes 1–3). It explains the structure and function of the nephron—the basic unit of function of the kidney (learning outcome 5), traces the pathway of blood through the blood vessels of the kidney, explains how glomerular filtrate is produced, and discusses the role of tubular reabsorption in urine production (learning outcomes 4, 6–8). It also describes the role of tubular secretion in urine formation (learning outcome 9) and discusses the structure of the ureters, urinary bladder, and urethra and how they function in micturition (learning outcomes 10 and 11).

A study of the urinary system is basic to understanding how the body maintains its chemistry within very narrow limits.

LEARNING OUTCOMES

After you have studied this chapter, you should be able to

17.1 Introduction

1. List the general functions of the organs of the urinary system.

17.2 Kidneys

2. Describe the locations and structure of the kidneys.
3. List the functions of the kidneys.
4. Trace the pathway of blood through the major vessels in a kidney.
5. Describe a nephron, and explain the functions of its major parts.

17.3 Urine Formation

6. Explain how glomerular filtrate is produced, and describe its composition.
7. Explain the factors that affect the rate of glomerular filtration and how this rate is regulated.
8. Discuss the role of tubular reabsorption in urine formation.
9. Define *tubular secretion*, and explain its role in urine formation.

17.4 Urine Elimination

10. Describe the structure of the ureters, urinary bladder, and urethra.
11. Explain the process and control of micturition.

FOCUS QUESTION

This afternoon, you attended a reception for the new president of the college. You enjoyed the salted nuts, smoked fish, cream puffs, petit fours, and other hors d'oeuvres that were served. How will the urinary system respond to this unusual food intake?

MASTERY TEST

Now take the mastery test. Do not guess. As soon as you complete the test, correct it. Note your successes and failures so that you can read the chapter to meet your learning needs.

1. The functions of the urinary system include the
   a. elimination of nitrogenous wastes and carbon dioxide.
   b. control of blood pressure.
   c. control of red blood cell production.
   d. regulation of pH.

2. The organ(s) of the urinary system that function(s) primarily to transport urine is (are) the
a. kidney.
b. urethra.
c. ureters.
d. bladder.

3. The kidneys are located
a. within the abdominal cavity.
b. between the twelfth thoracic and third lumbar vertebrae.
c. posterior to the parietal peritoneum.
d. just below the diaphragm.

4. The superior end of the ureters is expanded to form the funnel-shaped ______________________________.

5. Series of small elevations that project into the renal sinus and form the sinus wall are called
a. renal pyramids.
b. the renal medulla.
c. renal calyces.
d. renal papillae.

6. The blood supply to the nephron is via
a. the renal artery.
b. an interlobar artery.
c. an arcuate artery.
d. afferent arterioles.

7. The structure of the renal corpuscle consists of the
a. glomerulus.
b. glomerular capsule.
c. descending loop of Henle.
d. proximal convoluted tubule.

8. The transport mechanism used in the glomerulus is
a. filtration.
b. osmosis.
c. active transport.
d. diffusion.

9. High pressure in the glomerulus is maintained by the diameter of the ________________________.

10. The control of renin secretion is the function of the ________________ apparatus.

11. The end product of kidney function is ________________________________.

12. The fluid formed in the capillary cluster of the nephron is the same as blood plasma except for the absence of
a. glucose.
b. larger molecules of plasma protein.
c. bicarbonate ions.
d. creatinine.

13. Blood pressure affects urine formation because ____ of the blood is necessary to the transport mechanism used in the glomerulus.

14. How much fluid filters through the glomerulus in a 24-hour period?
a. 2 liters
b. 200 milliliters
c. 180 liters
d. 18 liters

15. Renin is secreted by the juxtaglomerular cells in response to a fall in
a. blood pressure.
b. sodium.
c. potassium.
d. fluid volume.

16. Renin is converted to angiotensin I in the ____________________ and to angiotensin II in the ________________.

17. Which of the following substances is (are) present in glomerular filtrate but not in urine?
a. urea
b. sodium
c. potassium
d. glucose

18. Substances such as sodium ions are reabsorbed in the
a. proximal convoluted tubule.
b. distal convoluted tubule.
c. descending limb of the loop of Henle.
d. ascending limb of the loop of Henle.

19. The permeability of the distal segment of the tubule to water is regulated by
a. blood pressure.
b. ADH.
c. aldosterone.
d. renin.

20. The mechanism by which greater amounts of a substance may be excreted in urine than were filtered from the plasma in the glomerulus is

a. tubular absorption.
b. active transport.
c. pinocytosis.
d. tubular secretion.

21. Which of the following substances enter(s) the urine via tubular secretion?

a. lactic acid
b. hydrogen ions
c. amino acids
d. potassium

22. The normal output of urine for an adult in an hour is

a. 20–30 mL.
b. 30–40 mL.
c. 40–50 mL.
d. 50–60 mL.

23. Urine is conveyed from the kidney to the bladder via the ______________________________.

24. Urine moves along the ureters via

a. hydrostatic pressure.
b. gravity.
c. peristalsis.

25. Inflammation of the bladder is called ______________________________.

26. The internal floor of the bladder has three openings in a triangular area called the ______________________________.

27. The third layer of the bladder is composed of smooth muscle fibers and is called the

a. micturition muscle.
b. detrusor muscle.
c. urinary muscle.
d. sympathetic muscle.

28. When stretch receptors in the bladder send impulses along parasympathetic paths, the individual experiences a sensation known as ______________________________.

29. The usual amount of urine voided at one time is about

a. 50 mL.
b. 150 mL.
c. 500 mL.
d. 1,000 mL.

30. Which of the following structures is under conscious control?

a. external urethral sphincter
b. internal urethral sphincter
c. bladder wall

STUDY ACTIVITIES

Aids to Understanding Words

Define the following word parts. (p. 468)

calyc-

cort-

detrus-

glom-

mict-

nephr-

papill-

trigon-

17.1 Introduction (p. 468)

List the functions of the urinary system.

17.2 Kidneys (pp. 468–472)

A. Identify the parts of the urinary system in the accompanying illustration (p. 468): kidney, bladder, ureter, aorta, inferior vena cava, hilum, renal artery, renal vein, urethra.

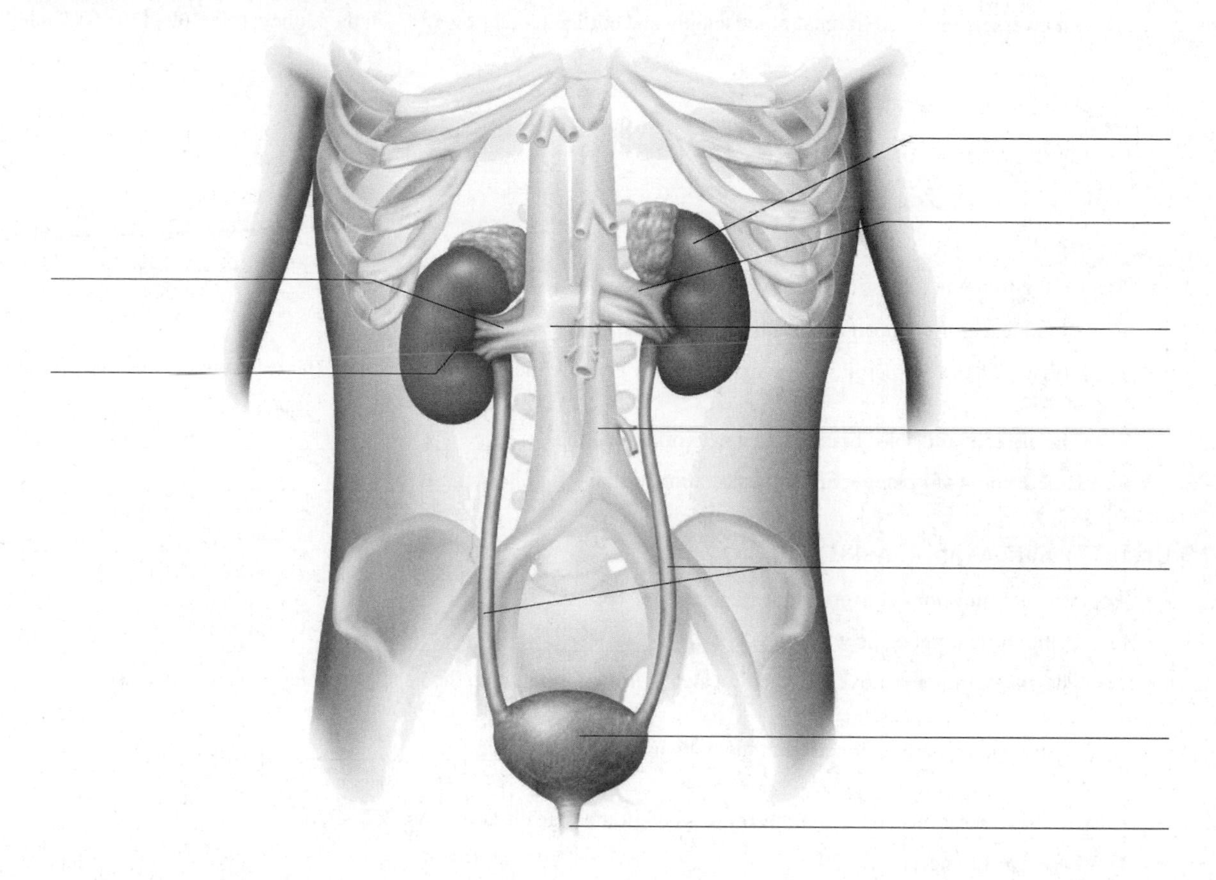

B. Describe the precise location of the kidneys. (p. 468)

C. Label these structures in the accompanying illustration (p. 469): renal pelvis, major and minor calyces, renal cortex, renal medulla, renal papilla, renal pyramid, ureter, renal column.

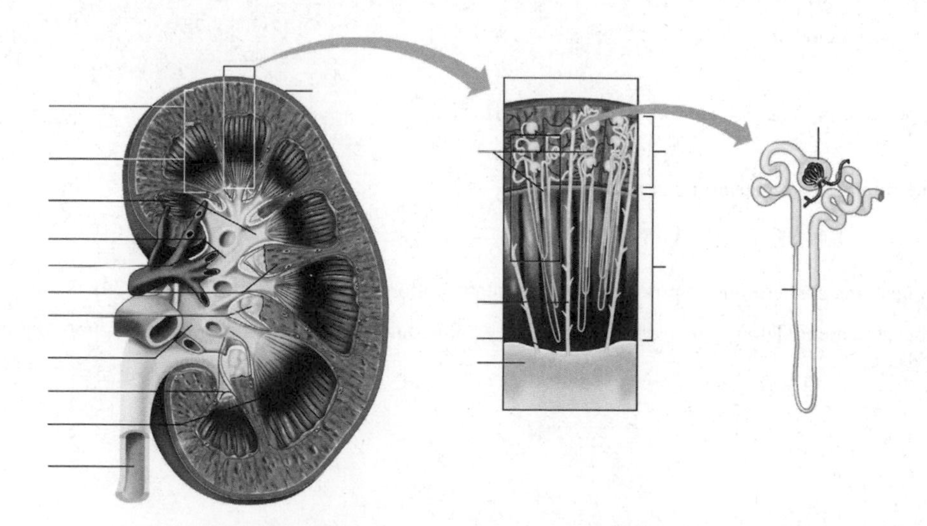

D. Describe the functions of the kidney and define the functional unit of the kidney. (p. 469)

E. Name the vessels involved in renal blood supply and outline blood flow through the kidney. (pp. 469–470)

F. Describe the structure of a nephron. (pp. 470–471)

G. Answer the following questions regarding the nephron's blood supply. (pp. 471–472)

1. Blood leaves the glomerulus via the ______________________________.
2. Pressure in the glomerulus is maintained partly because the efferent arteriole is (larger, smaller) than the afferent arteriole.
3. The efferent arterioles branch into a network called the ______________________________.
4. Blood from the kidney returns to circulation via the ______________________________.

17.3 Urine Formation (pp. 472–481)

A. Answer these questions concerning urine formation. (pp. 472–473)

1. Urine formation begins with ______________________________.
2. The excess water removed by the first step in urine formation to the venous circulation is returned using ______________________.
3. Hydrogen ions are removed from the body using ______________________________.

B. Answer these questions concerning the process of glomerular filtration. (pp. 473–476)

1. Describe glomerular filtration.
2. What factors influence the glomerular filtration rate?
3. What are the normal amount and composition of glomerular filtrate?
4. Why is renin secreted?
5. Describe the actions of angiotensin I and angiotensin II.
6. What is the role of atrial natriuretic peptide?

C. Answer these questions concerning the process of tubular reabsorption in the nephron. (pp. 477–479)

1. Explain this statement: Tubular reabsorption is selective. Illustrate it by describing the reabsorption of glucose and amino acids.

2. Define *renal plasma threshold.*

3. List the substances reabsorbed by the epithelium of the proximal convoluted tubule.

4. Describe water and sodium reabsorption in the proximal segment of the renal tubule.

D. Answer these questions concerning tubular secretion. (p. 479)

1. What is tubular secretion? Include what substances are secreted and how they are secreted.

2. Where are hydrogen ions secreted and why is this important?

3. How are potassium ions secreted and what is the consequence of their secretion?

E. Answer these questions concerning the regulation of urine concentration and volume. (p. 480)

1. What are the roles of ADH and aldosterone in urine concentration?

2. How are ADH and aldosterone secretion regulated?

3. What are the targets and results of ADH and aldosterone secretion?

4. What do urea and uric acid result from and how does the kidney process them?

F. Answer these questions regarding the composition of urine. (pp. 480–481)

1. What is the composition of normal urine?

2. What is the normal output of urine?

17.4 Urine Elimination (pp. 481–484)

A. Answer these questions concerning the ureters. (p. 481)

1. Describe the location and structure of the ureters.

2. How is urine moved through the ureters?

B. Answer these questions concerning the urinary bladder. (pp. 481–483)

1. How does the bladder change as it fills with urine?

2. How is urine prevented from spilling back into the ureters?

3. Describe the structure of the bladder wall.

C. Describe the process of micturition. Be sure to include both autonomic and voluntary events. (pp. 483–484)

D. Describe the urethra and outline the differences between the male and female urethra. (p. 484)

Clinical Focus Questions

A. Based on renal physiology, which electrolyte is most important to survival? Explain your answer.

B. List the possible reasons for finding red blood cells in urine.

C. What signs and symptoms would you expect to observe in an individual experiencing renal failure?

When you have finished the study activities to your satisfaction, retake the mastery test and compare your results with your initial attempt. If you are not satisfied with your performance, repeat the appropriate study activities.

CHAPTER 18

WATER, ELECTROLYTE, AND ACID-BASE BALANCE

OVERVIEW

This chapter presents the roles of several body systems in the maintenance of proper concentrations of water and electrolytes. It defines water and electrolyte balance and explains its importance (learning outcome 1). It tells how water enters the body, how water is distributed in various body compartments, and how it leaves the body (learning outcomes 2 and 3). It explains how electrolytes enter and leave the body and how their concentrations are regulated (learning outcome 4). Acid-base balance and the mechanisms regulating this balance are described, as well as the consequences of acid-base imbalance (learning outcomes 5–8).

Water and electrolyte balance affects and is affected by the various chemical interactions in the body. Knowledge of the mechanisms that control fluid and electrolyte balance and acid-base balance is essential to understanding the nature of internal environment.

LEARNING OUTCOMES

After you have studied this chapter, you should be able to

18.1 Introduction

1. Explain water and electrolyte balance.

18.2 Distribution of Body Fluids

2. Explain body fluid distribution in compartments.

18.3 Water Balance

3. List the routes by which water enters and leaves the body, and explain how water intake and output are regulated.

18.4 Electrolyte Balance

4. Explain how electrolytes enter and leave the body, and describe how electrolyte intake and output are regulated.

18.5 Acid-Base Balance

5. List the major sources of hydrogen ions in the body.
6. Distinguish between strong and weak acids and bases.
7. Explain how chemical buffer systems, the respiratory center, and the kidneys keep the pH of body fluids constant.

18.6 Acid-Base Imbalances

8. Describe the causes and consequences of an increase or decrease in body fluid pH.

FOCUS QUESTION

You have eaten a large bowl of salted popcorn while studying and suddenly realize you are very thirsty. Why are you thirsty?

MASTERY TEST

Now take the mastery test. Do not guess. As soon as you complete the test, correct it. Note your successes and failures so that you can read the chapter to meet your learning needs.

1. *Fluid and electrolyte balance* implies that the quantities of these substances entering the body ____________________ the quantities leaving the body.

2. Which of the following statements about fluid and electrolyte balance is (are) true?
a. Fluid balance is independent of electrolyte balance.
b. The concentration of an individual electrolyte is the same throughout the body.
c. Water and electrolytes occur in compartments in which the composition of fluid varies.
d. Water is evenly distributed throughout the tissues of the body.

3. More than 60% of body fluid occurs within the cells in a compartment called the ______________________ fluid compartment.

4. Blood and cerebrospinal fluid occur in the ______________________ fluid compartment.

5. Which of the following electrolytes is (are) most concentrated within the cell?
a. sodium
b. bicarbonate
c. chloride
d. potassium

6. Plasma leaves the capillary at the arteriole end and enters interstitial spaces because of ______________________ pressure.

7. Fluid returns from the interstitial spaces to the plasma at the venule end of the capillary because of ______________________ pressure.

8. The most important source of water for the normal adult is
a. beverages.
b. moist food, such as lettuce and tomatoes.
c. oxidative metabolism of nutrients.

9. Thirst is experienced when
a. the mucosa of the mouth begins to lose water.
b. salt concentration in the cells increases.
c. the hypothalamus is stimulated by increasing osmotic pressure of extracellular fluid.
d. the cortex of the brain is stimulated by shifts in the concentration of sodium.

10. The primary regulator of water output is
a. loss in the feces.
b. evaporation as sweat.
c. urine production.
d. loss with respiration.

11. The amount of water lost or retained is regulated by
a. changes in metabolic rate.
b. an increase or a decrease in respiratory rate.
c. the amount and consistency of feces.
d. an antidiuretic hormone.

12. The primary sources of electrolytes are ______________________ and ______________________.

13. List three routes by which electrolytes are lost.

14. Sodium and potassium ion concentrations are regulated by the kidneys and the hormone ______________________.

15. Calcium ion concentration in the plasma is regulated by ______________________.

16. Acid-base balance is mainly concerned with regulating ______________________ concentration.

17. Anaerobic respiration of glucose produces
a. carbonic acid.
b. lactic acid.
c. acetoacetic acid.
d. ketones.

18. The strength of an acid depends upon the
a. number of hydrogen ions in each molecule.
b. nature of the inorganic salt.
c. degree to which molecules ionize in water.
d. concentration of acid molecules.

19. A base is a substance that will ______________________ hydrogen ions.

20. A buffer is a substance that
a. returns an acid solution to neutral.
b. converts acid solutions to alkaline solutions.
c. converts strong acids or bases to weak acids or bases.
d. returns an alkaline solution to neutral.

21. The most important buffer system in plasma and intracellular fluid is the
a. bicarbonate buffer.
b. phosphate buffer.
c. protein buffer.

22. The respiratory center controls hydrogen ion concentration by controlling the ______________________ and ______________________ of respirations.

23. The slowest acting of the mechanisms that control pH is (are) the
a. buffer systems.
b. respiratory system.
c. kidneys.

24. Factors that increase the concentration of carbonic acid lead to ____________________ acidosis.

25. Factors that can lead to metabolic acidosis include
a. decreased glomerular filtration.
b. pneumonia.
c. diabetes mellitus.
d. vomiting and diarrhea.

STUDY ACTIVITIES

Aids to Understanding Words

Define the following word parts. (p. 490)

de-

extra-

im-

intra-

neutr-

18.1 Introduction (p. 490)

A. Define *water* and *electrolyte balance.*

B. Explain the interdependence of water and electrolyte balance.

18.2 Distribution of Body Fluids (pp. 490–492)

A. What are the different fluid compartments in the body and what percentage of the fluids does each contain? (p. 490)

B. What is the composition of extracellular fluid? (p. 491)

C. What is the composition of intracellular fluid? (p. 491)

D. What mechanisms are responsible for the movement of fluid and electrolytes from one compartment to another? Describe these mechanisms as fully as possible. (pp. 491–492)

18.3 Water Balance (pp. 492–493)

A. Answer these questions concerning water intake. (p. 492)

1. What is water balance?

2. What are sources of water intake?

B. Describe the mechanism that regulates the intake of water. (p. 492)

C. By what routes is water lost from the body? (p. 493)

D. The hormone that regulates water balance is ____________________________. How is this accomplished? (p. 493)

E. Describe the development and symptoms of dehydration. (p. 494)

18.4 Electrolyte Balance (pp. 493–497)

A. Answer these questions concerning electrolyte intake. (p. 493)

1. List the electrolytes that are important to cellular function.

2. What are the sources of these electrolytes?

B. List three routes by which electrolytes are lost. (p. 493)

C. Answer these questions concerning regulation of electrolyte balance. (pp. 493–497)

1. What is the role of aldosterone in electrolyte regulation?

2. What is the role of parathyroid hormone in electrolyte regulation?

3. What is the role of the kidney in electrolyte regulation?

4. How is the concentration of negatively charged ions regulated?

5. Describe the causes and symptoms of high and low concentrations of sodium and potassium.

18.5 Acid-Base Balance (pp. 497–499)

A. Answer these questions concerning acid-base balance. (p. 497)

1. What is an acid?

2. What is a base?

3. What is acid-base balance?

B. List and describe the sources of hydrogen ions in the body. (p. 497)

C. What is the difference between a strong acid or base and a weak acid or base? (p. 497)

D. Answer these questions concerning regulation of hydrogen ion concentration. (pp. 498–499)

1. In general, how is hydrogen ion concentration regulated by acid-base buffer systems?

2. Describe how hydrogen ion concentration works with the bicarbonate buffer system, the phosphate buffer system, and the protein buffer system.

3. How does the respiratory center regulate hydrogen ion concentration?

4. How do the kidneys help regulate hydrogen ion concentration?

5. How do these regulatory mechanisms differ from each other, especially with respect to speed of action?

18.6 Acid-Base Imbalances (pp. 500–502)

A. What is the normal blood pH value? What pH defines someone with acidosis?
What pH defines someone with alkalosis? (p. 500)

B. Compare metabolic and respiratory acidosis. (pp. 500–501)

C. Compare metabolic and respiratory alkalosis. (pp. 501–502)

Clinical Focus Questions

A. How is environmental temperature related to water and electrolyte balance?

B. How would you maintain water and electrolyte balance in hot weather for the following?

1. Infants

2. Young, athletic adults

3. Middle-aged, sedentary adults

4. Elderly adults who live alone

When you have finished the study activities to your satisfaction, retake the mastery test and compare your results with your initial attempt. If you are not satisfied with your performance, repeat the appropriate study activities.

UNIT 6
THE HUMAN LIFE CYCLE

CHAPTER 19

REPRODUCTIVE SYSTEMS

OVERVIEW

This chapter explains reproductive systems—unique systems because they are essential to the survival of the species rather than to the survival of the individual. The chapter explains the structure and functions of the male and female reproductive systems (learning outcomes 1–4, 6, 7). It also describes the hormonal events that control the male and female reproductive systems (learning outcomes 5 and 8), as well as the major events of the reproductive cycle (learning outcome 9). The structure of the mammary glands is described (learning outcome 10). The relative effectiveness of several methods of birth control is discussed (learning outcome 11). Lastly, the symptoms of sexually transmitted infections are listed (learning outcome 12).

Understanding the processes of sexual function contributes to an understanding of humans as sexual beings.

LEARNING OUTCOMES

After you have studied this chapter, you should be able to

19.1 Introduction

1. State the general functions of the male and female reproductive systems.

19.2 Organs of the Male Reproductive System

2. Describe the general functions of each part of the male reproductive system.
3. Outline the process of spermatogenesis.
4. Describe semen production and exit from the body.

19.3 Hormonal Control of Male Reproductive Functions

5. Explain how hormones control the activities of the male reproductive organs and the development of male secondary sex characteristics.

19.4 Organs of the Female Reproductive System

6. Describe the general functions of each part of the female reproductive system.
7. Outline the process of oogenesis.

19.5 Hormonal Control of Female Reproductive Functions

8. Explain how hormones control the activities of the female reproductive organs and the development of female secondary sex characteristics.
9. Describe the major events of a reproductive cycle.

19.6 Mammary Glands

10. Review the structure of the mammary glands.

19.7 Birth Control

11. Describe several methods of birth control, including the relative effectiveness of each method.

19.8 Sexually Transmitted Infections

12. List the general symptoms of sexually transmitted infections.

FOCUS QUESTION

How do the male and female reproductive systems ensure survival of the species?

MASTERY TEST

Now take the mastery test. Do not guess. As soon as you complete the test, correct it. Note your successes and failures so that you can read the chapter to meet your learning needs.

1. The primary sex organs (gonads) of the male reproductive system are the ________________________.

2. Male sex cells are produced by the ____ of the seminiferous tubules.

3. The two types of cells in the epithelium of the seminiferous tubules are ________________________, which support and nourish the ________________________, which give rise to sperm cells.

4. The enzyme-containing structure that helps the sperm cell penetrate the ovum is the
   a. acrosome.
   b. flagellum.
   c. zygote.
   d. spermatogonium.

5. Sex cells are produced in a process called ________________________.

6. How many chromosomes does each spermatogonium contain?

7. During meiosis, the chromosome number is
   a. reduced.
   b. increased.
   c. unchanged.

8. The nucleus in the head of the sperm contains ________________________ chromosomes.

9. The function of the epididymis is to
   a. produce sex hormones.
   b. provide the sperm with mobile tails.
   c. store sperm as they mature.
   d. supply some of the force needed for ejaculation.

10. Which of the following substances is (are) added to sperm cells by the seminal vesicle?
    a. acid
    b. fructose
    c. glucose

11. The function of the secretion of the bulbourethral glands is to
    a. neutralize the acid secretions of the vagina.
    b. nourish sperm cells.
    c. lubricate the penis.
    d. increase the volume of seminal fluid.

12. The process by which sperm cells become capable of fertilizing an ovum is ________________________.

13. Which of the following is not a male internal accessory reproductive organ?
    a. epididymis
    b. vas deferens
    c. testis
    d. seminal vesicle

14. The external organs of the male reproductive system include the
    a. penis.
    b. testes.
    c. prostate gland.
    d. scrotum.

15. Exposure to cold causes smooth muscle in the scrotum to (contract, relax).

16. Erection of the penis depends upon
    a. contraction of the perineal muscles.
    b. filling of the corpus spongiosum with arterial blood.
    c. enlargement of the glans penis.
    d. peristaltic contractions of the vas deferens.

17. Emission and ejaculation accompany orgasm.
    a. True
    b. False

18. Hormones that control male reproductive functions are secreted from the ________________________, the ________________________, and the ____.

19. The pituitary hormone that stimulates the testes to produce testosterone is
    a. gonadotropin-releasing hormone.
    b. FSH.
    c. LH (ICSH).
    d. ACTH.

20. In the male, growth of body hair, especially in the axilla, face, and pubis, and increased muscle and bone development are examples of ____ characteristics.

21. The hormone that stimulates the changes described in question 20 is ______________________.

22. The primary sex organs (gonads) of the female reproductive system are the ______________________.

23. In the primary sex organs of the female, the primary germinal epithelium is located
a. in the medulla.
b. between the medulla and cortex.
c. in the cortex.
d. on the free surface of the ovary.

24. How many mature egg cells are produced by each primary oocyte?

25. At puberty, the primary oocyte matures within the ______________________________________.

26. The egg is released from the ovary in a process called ______________________.

27. Which of the following statements is (are) true about the uterine (fallopian) tubes?
a. The end of the uterine tube near the ovary has many fingerlike projections called fimbriae.
b. The fimbriae are attached to the ovaries.
c. The inner layer of the ovarian tube is cuboidal epithelium.
d. There are cilia in the lining of the uterine tube that help move the egg toward the uterus.

28. The inner layer of the uterus is the ___________________________.

29. Which of the following statements about the vagina is (are) true?
a. The mucosal layer contains many mucous glands.
b. The bulbospongiosus muscle is primarily responsible for closing the vaginal orifice.
c. The vagina connects the uterus to the outer surface of the body.
d. The hymen is a membrane that covers the mouth of the cervix.

30. The organ of the female reproductive system that corresponds to the penis is the
a. vagina.
b. mons pubis.
c. clitoris.
d. labium major.

31. Which of the following tissues become(s) engorged and erect in response to sexual stimulation?
a. clitoris
b. labia minora
c. outer third of vagina
d. upper third of vagina

32. The hormonal mechanisms that control female reproductive functions are (more, less) complex than those in the male.

33. The primary female sex hormones are ______________________ and ______________________.

34. Which of the following secondary sex characteristics in the female seem(s) to be related to androgen concentration?
a. breast development
b. growth of axillary and pubic hair
c. female skeletal configuration
d. deposition of adipose tissue over hips, thighs, buttocks, and breasts

35. During the menstrual cycle, the event that seems to initiate ovulation is
a. increasing level of progesterone.
b. a sudden increase in concentration of LH.
c. decreasing level of estrogen.
d. a cessation of secretion of FSH.

36. After release of the egg, the follicle forms a(n) ____.

37. As the follicle develops, the level of which of the following hormones increases?
a. estrogen
b. progesterone
c. FSH
d. LH

38. As the hormone levels change in that part of the cycle before and immediately after ovulation, which of the following changes is seen in the uterus?
a. growth of the myometrium
b. increase in adipose cells of the perimetrium
c. thickening of the endometrium
d. decrease in uterine gland activity

39. The concentration of which of the following hormones decreases following ovulation?
a. estrogen
b. progesterone
c. FSH
d. LH

40. The cessation of the menstrual cycle in middle age is called ______________________.

41. Before puberty, the male and female breasts are (similar, dissimilar).

42. Which of the following contraceptive methods is not a barrier method?

a. condom
b. spermicide
c. contraceptive implant
d. cervical cap

43. Which of the following contraceptive methods also prevents transmission of sexually transmitted diseases?

a. contraceptive pill
b. diaphragm
c. condom
d. IUD

STUDY ACTIVITIES

Aids to Understanding Words

Define the following word parts. (p. 506)

andr-

ejacul-

fimb-

follic-

genesis-

labi-

mens-

mons-

puber-

19.1 Introduction (p. 506)

Describe the functions of the male and female reproductive systems.

19.2 Organs of the Male Reproductive System (pp. 506–513)

A. Answer the following questions regarding the male reproductive organs. (pp. 506–507)

1. Label these structures in the accompanying illustration (p. 507): urinary bladder, vas deferens, prostate gland, penis, urethra, prepuce, glans penis, epididymis, testis, scrotum, corpus cavernosum, corpus spongiosum, bulbourethral gland, ejaculatory duct, seminal vesicle, symphysis pubis, anus, ureter, large intestine.

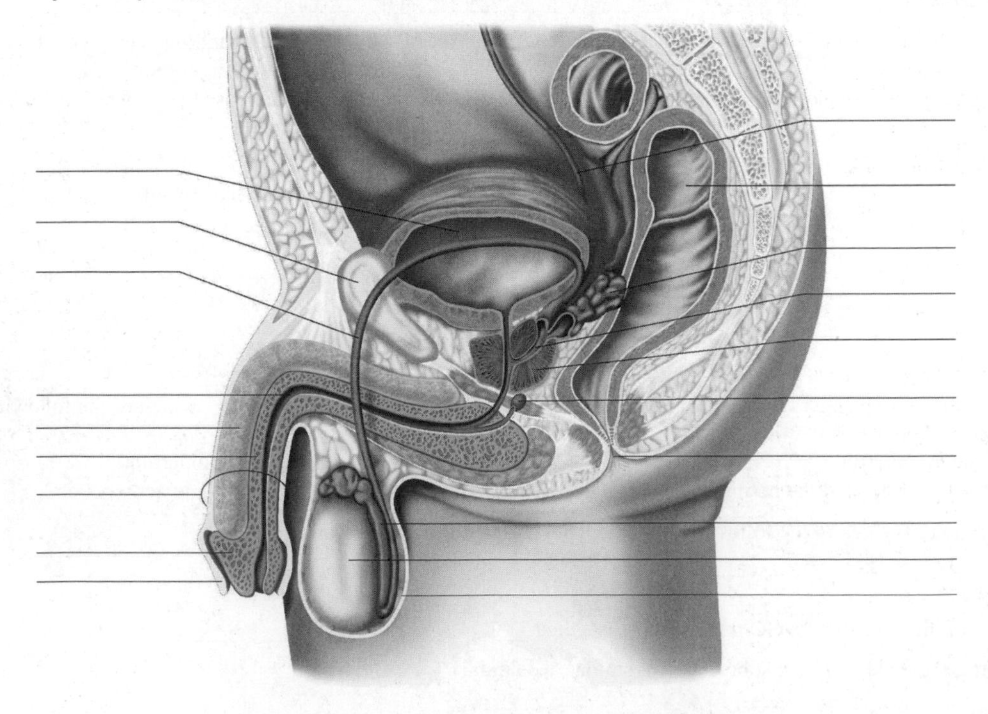

2. What is the primary function of the male reproductive system?

3. What are the primary organs of the male reproductive system?

B. Answer these questions concerning the structure of the testes. (p. 506)

1. What is the function of spermatogenic cells of the seminiferous tubules?

2. What is the function of interstitial cells?

3. In what cells does testicular cancer arise?

C. Name and describe the process and location of sperm formation. (p. 507)

D. Answer these questions concerning meiosis. (pp. 507–510)

1. What cells undergo meiosis?

2. Describe the process of meiosis in sperm.

3. Describe the structure of a sperm cell.

E. Answer the following questions about male internal reproductive organs. (pp. 510–511)

1. List the internal organs of the male reproductive system.

2. Describe the location and structure of the epididymis.

3. Describe the passage of the ductus deferentia through the pelvic cavity.

4. Answer these questions concerning the seminal vesicles.
   a. Where are the seminal vesicles located?

   b. What are the nature and function of the secretion of the seminal vesicles?

5. Answer these questions concerning the prostate gland.
   a. Where is the prostate gland located?

   b. What are the nature and function of the secretion of the prostate gland?

6. Describe the location and function of the bulbourethral glands (Cowper's glands).

7. Describe semen.

F. Answer these questions about male external reproductive organs. (p. 512)

1. What are the structure and the function of the scrotum?

2. Describe the structure of the penis.

G. Describe the events of erection, orgasm, and ejaculation. (pp. 512–513)

19.3 Hormonal Control of Male Reproductive Functions (pp. 513–515)

A. Answer these questions concerning hypothalamic and pituitary hormones. (p. 513)

1. What seems to initiate the changes that occur at puberty?

2. What are the functions of FSH and LH?

B. Answer these questions concerning male sex hormones. (pp. 513–515)

1. Where and when is testosterone produced?

2. What is the function of testosterone?

3. List the male secondary sex characteristics.

4. Describe the regulation of sex hormones in the male.

19.4 Organs of the Female Reproductive System (pp. 516–522)

A. Label these structures in the accompanying illustration (p. 516): ovary, uterine tube, uterus, vagina, anus, urinary bladder, urethra, clitoris, labium minor, labium major, symphysis pubis, fimbriae, cervix, rectum, vaginal orifice.

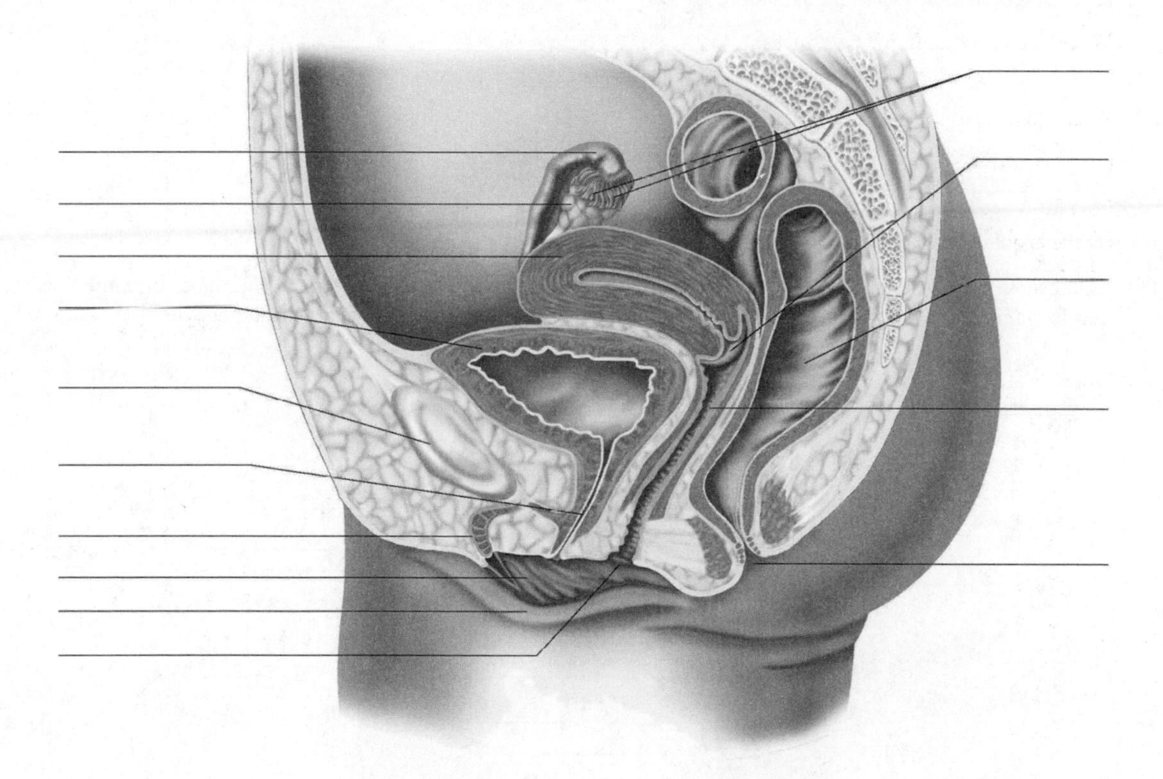

B. Answer the following questions regarding the ovaries. (pp. 516–517)

1. Describe the structure of the ovaries.

2. What are the primordial follicles?

C. Answer the following questions regarding oogenesis. (p. 517)

1. Describe the process of oogenesis.

2. How is this process different from spermatogenesis?

D. Answer these questions concerning the maturation of a follicle. (p. 518)

1. What stimulates maturation of a primary follicle at puberty?

2. What changes occur in the follicle as a result of maturation?

E. Answer these questions concerning ovulation. (p. 519)

1. What triggers ovulation?

2. What happens to the egg after it leaves the ovary?

F. Answer these questions about female internal reproductive organs. (pp. 519–521)

1. Label these structures in the accompanying illustration (p. 519): body of uterus, uterine tube, infundibulum, ovary, cervix, vagina, cervical orifice, fimbriae, follicle, endometrium, myometrium, perimetrium, oocyte.

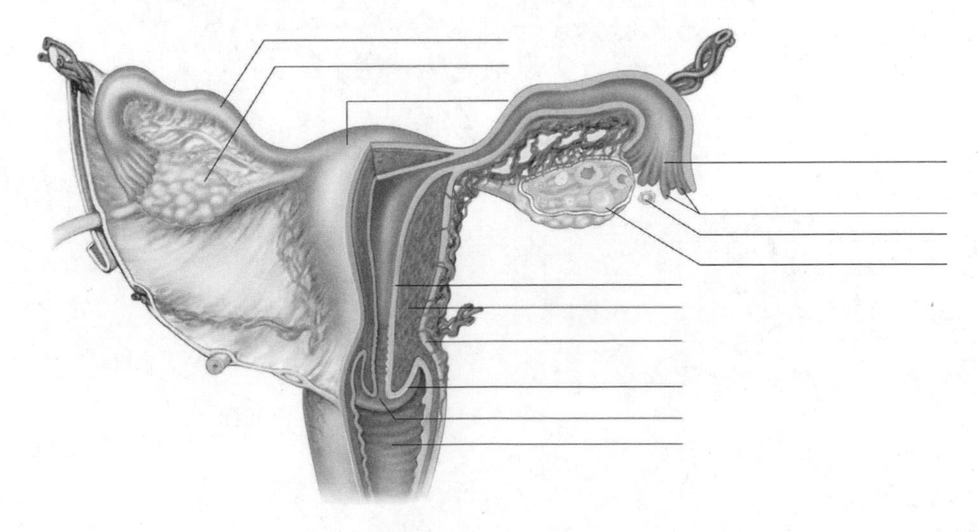

2. How does the structure of the uterine tube move the secondary oocyte toward the uterus?

3. Draw a uterus. Locate the body, fundus, cervix, endometrium, myometrium, and perimetrium and describe the function of this organ.

4. Describe the structure and function of the vagina.

G. Answer these questions about the female external reproductive organs. (p. 521)

1. Name the female external reproductive organs and tell what male external organs they correspond to.

2. Describe the labia majora and minora.

3. Describe the structure of the clitoris.

4. Describe the location of the vestibule and the function of the vestibular glands.

H. Describe the events of erection, lubrication, and orgasm in the female. (p. 521)

19.5 Hormonal Control of Female Reproductive Functions (pp. 522–524)

A. Answer these questions concerning female sex hormones. (p. 522)

1. What appears to initiate sexual maturation in the female?

2. What are the sources of female sex hormones?

3. What is the function of estrogen?

4. What is the function of progesterone?

5. What is the function of androgens?

B. Answer these questions concerning female reproductive cycles. (pp. 522–524)

1. How is a female's first menstrual cycle initiated?

2. Describe the events of the menstrual cycle. Include shifts in hormone levels, uterine changes, and ovarian changes.

3. Why do women athletes experience decreased or absent menstrual flow?

C. Answer these questions concerning menopause. (p. 524)

1. What is menopause?

2. What seems to be the cause of menopause?

19.6 Mammary Glands (pp. 525–527)

A. Describe the structure of the mammary glands. (p. 525)

B. What is the function of the mammary glands and where are they located? (p. 525)

C. What are the warning signs and treatments for breast cancer? (pp. 526–527)

19.7 Birth Control (pp. 526–530)

A. Describe the various mechanisms of contraception for each of the categories listed. (pp. 526–529)

1. Coitus interruptus
2. Rhythm method
3. Mechanical barriers
4. Chemical barriers
5. Combined hormone contraceptives
6. Injectable contraceptives
7. Intrauterine devices
8. Surgical methods

B. Outline any risks associated with these methods. (pp. 526–529)

19.8 Sexually Transmitted Infections (p. 530)

A. List the common symptoms of sexually transmitted infections (STIs).

B. What are the male and female complications of STIs?

Clinical Focus Question

Compare the various methods of contraception. How do moral and ethical factors influence an individual's contraceptive choices?

When you have finished the study activities to your satisfaction, retake the mastery test and compare your results with your initial attempt. If you are not satisfied with your performance, repeat the appropriate study activities.

CHAPTER 20

PREGNANCY, GROWTH, AND DEVELOPMENT

OVERVIEW

This chapter describes the events of fertilization and the changes that take place in the maternal body during pregnancy (learning outcomes 2, 3, and 6). It addresses the role of hormones in both the birth process and milk production (learning outcome 11). It also covers the events of embryonic and fetal development (learning outcomes 4, 5, 7–10) and the adjustments of the infant to extrauterine life (learning outcome 12). The concepts of growth and development are also covered (learning outcome 1). The chapter ends with a discussion on the processes of aging (learning outcome 13) as well as the mechanisms and modes of inheritance (learning outcomes 14 and 15).

LEARNING OUTCOMES

After you have studied this chapter, you should be able to

20.1 Introduction

1. Distinguish between growth and development.
2. Distinguish between the prenatal and postnatal periods.

20.2 Pregnancy

3. Describe fertilization.

20.3 Prenatal Period

4. List and provide details of the major events of cleavage.
5. Distinguish between an embryo and a fetus.
6. Discuss the hormonal changes in the maternal body during pregnancy.
7. List the structures produced by each of the primary germ layers.
8. Describe the major events of the embryonic stage of development.
9. Describe the major events of the fetal stage of development.
10. Trace the general path of blood through the fetal cardiovascular system.
11. Explain the role of hormones in the birth process and milk production.

20.4 Postnatal Period

12. Describe the major physiological adjustments required of the newborn.

20.5 Aging

13. Distinguish between passive aging and active aging.

20.6 Genetics

14. Distinguish among the modes of inheritance.
15. Describe the components of multifactorial traits.

FOCUS QUESTION

How is a unique individual produced from the union of two cells?

MASTERY TEST

Now take the mastery test. Do not guess. As soon as you complete the test, correct it. Note your successes and failures so that you can read the chapter to meet your learning needs.

1. An increase in size and number of cells is referred to as ______________________________.

2. Fertilization takes place in the
a. vagina.
b. cervix.
c. uterus.
d. uterine tube.

3. During sexual intercourse, semen containing sperm cells are deposited near the ________________________.

4. Which of the following is thought to be the mechanism by which the sperm enters the egg?
a. An antigen-antibody reaction briefly alters the cell membrane of the egg.
b. The structure of the cell membrane of the egg allows entry.
c. The head of the sperm has an enzyme that permits digestion through the membrane of the egg.
d. The mechanism is unknown.

5. Female infertility can be due to
a. hypersecretion of pituitary gonadotropic hormones.
b. endometriosis.
c. scarring of the uterine tubes.
d. mucous plug in the cervix secondary to infection.

6. The first phase in embryonic development is called ________________________.

7. Implantation takes place by the end of ________________________ week(s) following fertilization.

8. Following cleavage, the fertilized egg forms a sixteen-cell ball called the________________________.

9. The fertilized egg that begins to attach to the endometrium of the uterus is known as the________________________.

10. Implantation begins the stage of development known as the ________________________ stage.

11. After the eighth week of pregnancy, the products of conception are known as the________________________.

12. The hormone that maintains the corpus luteum following implantation is
a. LH.
b. HCG.
c. progesterone.
d. FSH.

13. The primary source of hormones needed to support a pregnancy after the first three months is the ________________________.

14. Breast development during pregnancy is stimulated by ________________________.

15. The ectoderm, mesoderm, and endoderm are ________________________ layers.

16. From which of the layers of the embryonic disc do the hair, nails, and glands of the skin arise?
a. endoderm
b. ectoderm
c. mesoderm

17. Which of the following structures arises from the mesoderm?
a. lining of the mouth
b. muscle
c. lining of the respiratory tract
d. epidermis

18. At what time does the embryonic disc become a cylinder?
a. four weeks of development
b. six weeks of development
c. eight weeks of development
d. four days of development

19. The chorion in contact with the endometrium becomes the ________________________.

20. The membrane covering the embryo is called the ________________________.

21. The vessels in the umbilical cord are
a. one artery and one vein.
b. one artery and two veins.
c. two arteries and one vein.
d. two arteries and two veins.

22. In the embryo, blood cells are formed in the
a. amnion.
b. placenta.
c. allantois.
d. yolk sac.

23. A factor that damages the embryo during a period of rapid growth is a(n) ________________________.

24. The embryonic stage ends at ________________________ weeks.

25. Pregnancy can be diagnosed ten days after fertilization by testing the urine of the woman for which of the following substances?
a. estrogen
b. progesterone
c. human chorionic gonadotropin
d. amniotic fluid

26. At the beginning of the fetal stage of development, (most, few) of the body structures are formed.

27. Skeletal muscles become active in the ________________________ lunar month.

28. A fetus is full-term at the end of the ____________________________ lunar month.

29. Oxygen- and nutrient-rich blood reach the fetus from the placenta via the umbilical ________________________.

30. The ductus venosus shunts blood around the
a. liver.
b. spleen.
c. pancreas.
d. small intestine.

31. The structures that allow blood to avoid the nonfunctioning fetal lungs are the ________________________ and the ____.

32. Labor is initiated by a decrease in ______________________ levels and the secretion of ________________________ by the posterior pituitary gland.

33. Bleeding following expulsion of the afterbirth is controlled by
a. hormonal mechanisms.
b. increased fibrinogen levels.
c. contraction of the uterine muscles.
d. sympathetic stimulation of arterioles.

34. Milk production is stimulated by the hormone ________________________.

35. Milk is secreted from the breast
a. as production fills the duct structure.
b. in response to hormonal stimulation.
c. in response to mechanical stimulation of the nipple (i.e., sucking).
d. in response to gonadotropins.

36. The factor that decreases the effort required for an infant to breathe after the first breath is ____________________.

37. The primary energy source for the newborn is
a. glucose.
b. fat.
c. protein.

38. An infant's urine is (more, less) concentrated than an adult's.

39. Which of the following fetal structures close(s) as a result of a change in pressure?
a. ductus venosus
b. ductus arteriosus
c. umbilical vessels
d. foramen ovale

40. The science that studies why individuals vary in terms of characteristics is ________________.

STUDY ACTIVITIES

Aids to Understanding Words

Define the following word parts. (p. 537)

allant-

chorio-

cleav-

hetero-

hom-

lacun-

morul-

nat-

troph-

umbil-

20.1 Introduction (p. 537)

Define *growth* and *development*.

20.2 Pregnancy (pp. 537–540)

A. Define *pregnancy*. (p. 537)

B. Describe how the gametes (sex cells) are transported. (p. 537)

C. Describe the process of fertilization. (pp. 538–539)

D. What are some of the causes of female infertility? (p. 538)

20.3 Prenatal Period (pp. 541–553)

A. Describe the events of pregnancy from fertilization to implantation. (pp. 541–542)

B. Describe the hormonal changes that occur during pregnancy. (pp. 542–543)

C. Answer these questions concerning the embryonic stage. (pp. 543–547)

1. At what point does the embryonic stage begin and how long does the embryonic stage last?

2. What is the embryonic disc?

3. At what point do the three primary germ layers appear and what is the embryo called at this point?

4. List the structures that arise from the primary germ layers.

   ectoderm

   mesoderm

   endoderm

5. What is the chorion and how does it develop?

6. Describe how the fetal blood supply is established and the formation of the umbilical vessels.

7. What is the function of the allantois and the yolk sac?

8. What are teratogens and how do they affect the developing embryo?

D. Describe the major events that occur during the fetal development stage. (pp. 547–551)

Month	Major events of growth and development
Third month	
Fourth month	
Fifth month	
Sixth month	
Seventh month	

Eighth month

Ninth month

E. Answer the following questions concerning fetal blood and circulation. (pp. 549–551)

1. Outline the route of blood in fetal circulation.

2. Describe why fetal hemoglobin is different.

3. What are the differences between the fetal and postnatal cardiovascular systems?

F. Answer these questions concerning the birth process. (pp. 551–552)

1. What is the birth process and how is it initiated?

2. What stimulates secretion of oxytocin and what is its role?

3. How is the secretion of oxytocin regulated?

4. What is the afterbirth?

5. How is bleeding controlled once the placenta is expelled?

G. Describe the production and secretion of milk. (pp. 552—553)

20.4 Postnatal Period (pp. 553–555)

A. Fill in the following information concerning changes that occur in the neonatal period to allow the newborn to adjust to extrauterine life. (pp. 553–555)

1. Respiration

2. Nutrition

3. Urine formation

4. Temperature control

5. Circulation

20.5 Aging (pp. 555–556)

A. Describe the process of passive aging. (pp. 555–556)

B. Describe the process of active aging. (p. 556)

20.6 Genetics (pp. 556–560)

A. Answer the following questions concerning chromosomes and genes. (pp. 556–557)

1. What is the field of genetics?
2. Inherited traits are determined by DNA sequences called ____________.
3. The environment can influence the way genes are expressed. (True/False) __________
4. What are autosomes? What are the sex chromosomes?
5. A way of displaying the 23 chromosome pairs is called a
   a. genetic map.
   b. karyotype.
   c. genetic code.
   d. phenotype.
6. Variant forms of a gene are called _____________.
7. The combination of alleles for a gene constitutes an individual's ____________. The appearance of a health condition associated with a particular combination of alleles is the individual's ____________.

B. Answer the following questions concerning the modes of inheritance. (pp. 557–560)

1. Describe an autosomal recessive inheritance pattern.

2. What is a pedigree?

3. Describe an autosomal dominant inheritance pattern.

4. What is X-linked inheritance and why do sons always inherit traits from their mothers?

C. What are multifactorial traits? (p. 560)

Clinical Focus Question

What advice would you give to a young woman who is considering becoming pregnant? How would the woman's age, education, family history, and economic status influence your answer?

When you have finished the study activities to your satisfaction, retake the mastery test and compare your results with your initial attempt. If you are not satisfied with your performance, repeat the appropriate study activities.

MASTERY TEST ANSWERS

1 Mastery Test Answers

1. c
2. b
3. c
4. Latin and Greek
5. anatomy
6. physiology
7. always
8. b
9. atoms, molecules, cell, tissue, organ, organ system
10. metabolism
11. movement, responsiveness, growth, reproduction, respiration, digestion, absorption, circulation, assimilation, excretion
12. water
13. energy, living matter
14. energy
15. increases
16. hydrostatic
17. a
18. 1. d; 2. b; 3. c
19. axial portion
20. appendicular
21. cranial, vertebral, thoracic, abdominopelvic
22. diaphragm
23. mediastinum
24. a, b, c
25. oral, nasal, orbital, middle ear
26. pleural cavity
27. pericardial
28. abdominopelvic
29. integumentary
30. a-3, b-4, c-2, d-2, e-4, f-2, g-2, h-3, i-2, j-1
31. b, c
32. c
33. b
34. body regions

2 Mastery Test Answers

1. a
2. Matter is anything that has weight and takes up space. It occurs in the forms of solids, liquids, and gases.
3. elements

4. carbon, hydrogen, oxygen, and nitrogen
5. e
6. a-3, b-1, c-2
7. protons
8. a
9. c
10. protons and neutrons
11. The outer shell of its atoms has its maximum number of electrons.
12. a
13. a
14. a
15. polar covalent
16. different
17. molecular
18. structural
19. synthesis, decomposition
20. reversible
21. catalyst
22. acid
23. bases
24. hydrogen ions
25. 7
26. electrolyte
27. a. I, b. O, c. O, d. I, e. O, f. O
28. carbon, hydrogen, oxygen
29. fatty acids, glycerol
30. energy
31. lipids
32. protein
33. nitrogen
34. its conformation or structure
35. RNA—ribonucleic acid, DNA—deoxyribonucleic acid
36. a

3 Mastery Test Answers

1. numbers, types; shapes and organelles
2. protective
3. a, b
4. cytoplasm, nucleus
5. b
6. signal transduction
7. selectively permeable
8. b

9. water
10. trans-membrane, integral
11. sodium, potassium
12. endoplasmic reticulum
13. a
14. outside
15. energy
16. b
17. liver, kidneys
18. movement
19. d
20. b
21. nucleolus, chromatin
22. cellular energy
23. diffusion
24. facilitated diffusion
25. osmosis
26. a
27. filtration
28. active transport
29. pinocytosis
30. phagocytosis
31. b
32. DNA replication
33. a-2, b-4, c-1, d-3
34. differentiation

4 Mastery Test Answers

1. c
2. anabolic metabolism
3. catabolic metabolism
4. a
5. c
6. protein
7. hydrolysis
8. enzyme
9. c
10. b
11. b
12. ase
13. Shape; conformation
14. enzyme, substrate,
15. coenzyme

16. a
17. oxidation
18. anaerobic respiration, glycolysis
19. aerobic respiration
20. oxygen
21. ATP
22. metabolic pathway
23. a, c
24. DNA (deoxyribonucleic acid)
25. adenine, thymine, cytosine, guanine
26. nucleus, cytoplasm
27. messenger, transfer
28. uracil
29. c
30. mutation
31. rate-limiting enzyme
32. d
33. interphase

5 Mastery Test Answers

1. epithelial, connective, muscle, and nervous tissue
2. similar
3. b
4. a, c, d
5. basement membrane
6. a-4, b-1, c-5, d-2, e-3
7. transitional epithelium
8. exocrine
9. exocytosis
10. a, c, d
11. fixed
12. c
13. collagen
14. elastin
15. fibroblasts
16. a,c,b,d
17. hyaline
18. fibrocartilage
19. slowly
20. bone
21. plasma
22. smooth, skeletal, cardiac
23. nervous

24. serous, mucous, cutaneous, synovial
25. b
26. b

6 Mastery Test Answers

1. epidermis
2. dermis
3. subcutaneous layer or hypodermis
4. b
5. a
6. equal
7. dermis
8. arrector pili muscles
9. c
10. keratinization
11. a, c
12. sweat
13. melanoma
14. b
15. c
16. a
17. redness, heat, swelling, and pain
18. deep

7 Mastery Test Answers

1. b
2. b
3. d
4. compact
5. spongy or cancellous
6. marrow
7. intramembranous bones
8. the patella
9. endochondral bones
10. d
11. osteocytes; osteoclasts
12. is remodeled
13. a, c
14. nutrition, hormone secretions, physical exercise
15. cartilaginous callus
16. levers
17. a, c

18. c
19. a, c
20. skull, hyoid, vertebral column, thoracic cage
21. pectoral girdle, arms or upper limbs, pelvic girdle, legs or lower limbs
22. c
23. b
24. occipital
25. maxillary
26. fontanels
27. b
28. c
29. a, b, c, d
30. transverse process and body, sternum
31. clavicles, scapulae
32. radius
33. a
34. a
35. c
36. b
37. c
38. b
39. a.
40. b.

8 Mastery Test Answers

1. a
2. skeletal muscle tissue, blood, nervous tissue, connective tissue
3. fascia
4. b
5. actin, myosin
6. c
7. mild muscle strain
8. actin filaments
9. c
10. motor unit
11. a
12. acetylcholine
13. adenosine triphosphate (or ATP)
14. creatine phosphate
15. b
16. a, c, d
17. lactic acid
18. muscle

19. threshold stimulus
20. c
21. a
22. a
23. decrease
24. multiunit, visceral
25. a, c
26. more slowly
27. rapidly
28. orgin; insertion
29. antagonists
30. d
31. a
32. b
33. linea alba
34. b

9 Mastery Test Answers

1. neuron
2. d
3. a
4. b
5. c
6. sensory, integrative, motor
7. somatic
8. neuroglial
9. neuroglial
10. capillaries, astrocytes
11. cell body, dendrites, axons
12. axonal hillock
13. c
14. bipolar, unipolar, multipolar
15. sensory, interneuron, motor
16. sodium
17. permeability
18. action potential
19. larger, smaller
20. synapse
21. neurotransmitters
22. a
23. processing
24. c
25. nerve

26. reflex
27. a
28. c
29. c
30. b
31. cerebrum, diencephalon, brain stem, cerebellum
32. a
33. a-3, b-4, c-3, d-1, e-2, f-1
34. left
35. choroid plexuses
36. d
37. b
38. limbic system
39. reticular formation
40. cerebellum
41. somatic, autonomic
42. 12, brain stem
43. 2, 6
44. 31
45. autonomic
46. lateral horn, thoracic, lumbar
47. a, b
48. c, d

10 Mastery Test Answers

1. chemoreceptors—change in concentration of chemicals; pain receptors—tissue damage; thermoreceptors—change in temperature; mechanoreceptors—change in pressure or movement; photoreceptors—light energy
2. c
3. sensory adaptation
4. skin, muscles, joints, viscera
5. a
6. c
7. a, c
8. acute fibers
9. reticular formation
10. endorphins
11. smell, taste, hearing, equilibrium (static, dynamic), sight
12. b
13. within the nasal cavity
14. olfactory nerve or tracts
15. c
16. b

17. sweet, salty, sour, bitter umami
18. equilibrium
19. c, d
20. eustachian tube (auditory tube)
21. osseous labyrinth, membranous labyrinth
22. vestibule, semicircular canals, cochlea
23. a
24. a
25. c
26. cornea
27. a
28. b
29. optic nerve
30. a
31. aqueous humor
32. pupil
33. retina
34. b
35. refraction
36. rods, cones
37. a-1, b-2, c-1, d-2
38. rhodopsin, opsin, retinal
39. optic chiasma

11 Mastery Test Answers

1. phermones
2. hormone
3. paracrine; autocrine
4. exocrine
5. endocrine `
6. plasma membrane
7. b
8. prostaglandins
9. a, b
10. hypothalamus
11. hypothalamus
12. a, b
13. a, b, c
14. prolactin
15. a, d
16. c
17. a
18. thyroxine, triiodothyronine

19. c, d
20. iodine
21. calcitonin
22. b, d
23. a, b, c
24. a,
25. epinephrine, norepinephrine
26. a
27. b, c
28. male
29. islets of Langerhans or pancreatic islets
30. glucagon
31. b, d
32. Type II
33. b
34. infection
35. a
36. d

12 Mastery Test Answers

1. connective
2. 55
3. b
4. a, b, c, d
5. 4,600,000–6,200,000; 4,200,000–5,400,000
6. a
7. 90–120 days
8. erythropoietin
9. b
10. a, c
11. iron
12. a
13. b
14. interleukins; colony stimulating factors
15. a
16. b
17. 5,000, 10,000
18. a, c
19. histamine, heparin
20. thrombopoietin
21. a-1, b-1, c-2, d-2, e-3
22. oxygen, carbon dioxide, nitrogen, nitrogen has no physiologic function
23. amino acids, simple sugars, nucleotides, lipids, vitamins, minerals

24. b, d
25. b
26. fibrinogen, fibrin
27. d
28. positive feedback
29. embolus
30. b
31. a
32. c

13 Mastery Test Answers

1. mediastinum
2. a
3. a
4. pericarditis
5. epicardum, myocardium, endocardium
6. b
7. atria; ventricles
8. a, d
9. b
10. chordac tendinae
11. coronary arteries
12. cardiac cycle
13. c
14. b
15. a
16. electrocardiogram
17. a
18. decrease
19. b, c ,d
20. vasoconstriction
21. plaque, atherosclerosis
22. c
23. b, d
24. filtration, osmosis, diffusion
25. a
26. a
27. b
28. valves
29. b
30. stroke volume
31. heart action, blood volume, viscosity, peripheral resistance
32. c

33. parasympathetic
34. b
35. a
36. d
37. left atrium
38. brachiocephalic artery, common carotid artery, left subclavian artery
39. common iliac

14 Mastery Test Answers

1. lymphatic
2. lymphatic capillaries, collecting ducts
3. b
4. b
5. c
6. a, c
7. veins
8. edema
9. c
10. d
11. b
12. thymosin, T
13. spleen
14. b, d
15. pathogens
16. c
17. a, d
18. redness, swelling, heat, pain
19. c
20. adaptive immunity
21. thymus gland
22. antigens
23. d
24. cell-mediated immunity
25. d
26. G, A, M
27. complement
28. accessory cell
29. d
30. passive
31. b
32. helper T cells, B cells, cytotoxic T cells
33. tissue rejection reaction
34. autoimmunity

15 Mastery Test Answers

1. a
2. alimentary canal
3. accessory organs
4. a
5. d
6. mixing, propelling
7. no
8. lingual frenulum
9. c
10. a
11. a, b
12. d
13. increase
14. b
15. c
16. peristalsis
17. b
18. c
19. b
20. vitamin B_{12}
21. a, b, d
22. a, d
23. inhibits
24. chyme
25. a
26. a
27. a
28. b, c, d
29. alkaline
30. upper right
31. d
32. a
33. d
34. Kuppfer
35. bile salts
36. a, d,
37. cholecystokinin
38. a, b, d
39. duodenum, jejunum, ileum
40. b
41. most
42. are

43. diarrhea
44. cecum
45. b
46. electrolytes, water
47. a, c
48. water
49. essential nutrients
50. a, b
51. cellulose
52. b, d
53. d
54. linoleic acid or linolenic acid
55. cholesterol
56. a, d
57. amino acids
58. complete
59. C
60. D
61. calcium, phosphorus
62. oxygen

16 Mastery Test Answers

1. respiration
2. a-1, b-2, c-1, d-2, e-3
3. nasal cavity, larynx
4. b, c
5. a
6. c
7. alveolar ducts
8. visceral pleura
9. parietal pleura
10. larger
11. b
12. contracts, increasing, decreasing
13. c
14. pleural
15. surfactant
16. emphysema
17. b, c, d
18. d
19. medulla, pons
20. c
21. pontine

22. a
23. carbon dioxide
24. alveolus, capillary
25. pressure
26. partial pressure
27. hemoglobin
28. c

17 Mastery Test Answers

1. b, c, d
2. b, c
3. b, c
4. renal pelvis
5. d
6. d
7. a, b
8. a
9. efferent arteriole
10. juxtaglomerular
11. urine
12. b
13. hydrostatic pressure
14. c
15. a, b,c
16. bloodstream, lungs
17. d
18. a, d
19. b
20. d
21. b, d
22. d
23. ureters
24. c
25. cystitis
26. trigone
27. b
28. urgency
29. b
30. a

18 Mastery Test Answers

1. equal
2. c
3. intracellular
4. extracellular
5. d
6. hydrostatic
7. osmotic
8. a
9. c
10. c
11. d
12. food, beverages
13. perspiration, feces, urine
14. aldosterone
15. parathyroid hormone
16. hydrogen ion
17. b
18. c
19. combine with
20. c
21. c
22. rate, depth
23. c
24. respiratory
25. a, c, d

19 Mastery Test Answers

1. testes
2. spermatogenic cells
3. supporting cells, spermatogenic cells
4. a
5. meiosis
6. 46
7. a
8. 23
9. c
10. b
11. c
12. capacitation
13. c
14. a, d

15. contract
16. b
17. a
18. testes, hypothalamus, anterior pituitary gland
19. c
20. secondary sexual
21. testosterone
22. ovaries
23. d
24. 1
25. primary follicle
26. ovulation
27. a, d
28. endometrium
29. b, c
30. c
31. a, c
32. more
33. estrogen, progesterone
34. b, c
35. b
36. corpus luteum
37. a, b
38. c
39. c, d
40. menopause
41. similar
42. b, c
43. c

20 Mastery Test Answers

1. growth
2. d
3. cervix
4. c
5. a, b, c, d
6. cleavage
7. one
8. morula
9. blastocyst
10. embryonic

11. fetus
12. b
13. placenta
14. placental lactogen
15. primary germ
16. b
17. b
18. a
19. placenta
20. amnion
21. c
22. c, d
23. teratogen
24. 8
25. c
26. most
27. fifth
28. tenth
29. vein
30. a
31. foramen ovale, ductus arteriosus
32. progesterone, oxytocin
33. c
34. prolactin
35. c
36. surfactant
37. b
38. less
39. d
40. genetics

Notes

Notes